Science, Paradox, and the Moebius Principle

The Evolution of a "Transcultural" Approach to Wholeness

by
Steven M. Rosen

State University of New York Press

Published by
State University of New York Press, Albany

© 1994 State University of New York

For information, address the State University of New York Press,
State University Plaza, Albany, NY 12246

Library of Congress Cataloging-in-Publication Data

Rosen, Steven M.
    Science, paradox, and the Moebius principle : the evolution of a
  "transcultural" approach to wholeness / by Steven M. Rosen.
      p.   cm. — (SUNY series in science, technology, and society)
  Includes bibliographical references and index.
    ISBN 0-7914-1769-7 (acid-free). — ISBN 0-7914-1770-0 (pbk. : acid
 -free
    1. Science—Philosophy. 2. Knowledge, Theory of.
 3. Bohm, David.   I. Title. II. Series.
 Q175.R547   1994
 510—dc20   93-3091
                                                      CIP

10 9 8 7 6 5 4 3 2 1

To Marlene

CONTENTS

vii

FOREWORD

Dichotomies seem to hold a fascination for Steven Rosen, driving him to find ways to integrate such troubling ones in the world of physics as locality/nonlocality, symmetry/asymmetry, continuity/discontinuity, infinity/finitude, and, ultimately, the mind/body duality flowing directly from our Cartesian heritage. This volume represents the evolution of his thought as he seeks to come to terms with the paradoxes that beset both science and philosophy. In his own words, his goal is to "confront crises and fragmentation in contemporary thought by offering a concrete intuition of thoroughgoing wholeness." In this effort and to his credit, he has embraced rather than backed away from the data of parapsychological research. He has sought (with considerable success, I think) a way of working toward a worldview that is inclusive rather than reductive, that embodies rather than avoids the counterintuitive markers that betray mysteries still to be solved. An extraordinarily successful strategy in achieving these ends is his use of the Moebius principle as exemplified in the geometry of the Moebius strip, the Necker cube, the Klein bottle, and the unique perspective of the paintings of M. C. Escher. All are used very effectively to capture the essence of his thesis.

Perhaps I can give the reader some idea of the feeling this book left me with by offering an up-to-date version of the "emperor's new clothes." In this story, the protagonist is not a little boy (although he does possess that childlike sense of wonder that can't be satisfied by anything less than the truth) but a very knowledgeable observer standing on the shoulders of such giants as Einstein, Heisenberg, Bohm, Whitehead, Heidegger, and Jung. From that perch, he views the "emperor" (our current view of reality) proudly displaying the clothes of the New Physics. With the exception of our observer, the crowd gasps in awe at the elegant and pragmatic formalisms of the quantum mechanical outfit the "emperor" is wearing. Certainly they are finer than any Newtonian tailor could have woven with those old-fashioned ideas of time, space, and causality. What catches the eye of this observer is how subtly the dualisms intrinsic to the old threads are woven into the garment of the new. One by one these threads are recognized for what they are, abstractions created to maintain a formalism that works (the "emperor" is quite content with it) but leaves visible all that doesn't fit (nonlocality, the intrusion of discontinuity into the continuum of space, the ever-intruding infinities).

To accommodate these intrusive findings, the author arrives at the concept of nondual duality (or alternatively, a dualistic monism). This is an integrative mode that regards each aspect of the various dualities as opposing features of an ongoing dialectic. Both poles of the various paradoxes are given their just due as each emerges as an aspect of a single totality. To achieve this end, he goes about carefully disassembling the logico-empirical methodology that underpins current science and, in so doing, reveals the various ways it tends to conceal the intrusion of *psyche* into the world of objective reality. One cannot, without getting into serious theoretical difficulties, weave a garment made exclusively of matter (*physis*) or spirit (*psyche*).

This volume holds a special appeal to me as a psychiatrist, for what it has contributed to my understanding of two special interests of mine, the nature of dreaming consciousness and the challenge of the "paranormal." As a clinician and as an experimentalist, I feel a special affinity to dreams, being drawn to them not only because of their healing potential but also because of the way they occasionally reach across space to reflect unknowable events that occur at a distance (telepathy) and across time to bring us non-inferential knowledge of future events (precognition). My interest in parapsychology goes back to early personal experiences, my encounter with telepathic dreams of patients in the course of my work as a psychoanalyst, and finally, the experimental studies I engaged in to establish the reality of the paranormal dream under laboratory conditions.

These interests resulted in my looking beyond psychology for theoretical leverage with which to confront the intrinsic mystery of dreaming consciousness and its connection to the paranormal. Physicists are faced with their own puzzles, which seem to be a permanent feature of quantum theory. Might the convergence of these mysteries provide clues that could move us beyond the currently accepted limiting parameters of reality? As a layman pursuing these interests, I have read many excellent books that try to convey the basic concepts of quantum mechanics to the general public. Unlike the present volume, they all seem to stop short just at the point when the mystery deepens. As Richard Feynman notes with regard to the odd behaviour of the electron in the double-slit experiment, "We say 'at the present time' but we suspect very strongly that it is something that will be with us forever—that it is impossible for us to beat that puzzle—that is the way nature really is." Dr. Rosen's motto seems to be *plus ultra*; and in his search for the "more beyond," he has launched a broad theoretical project designed to cast a new light on some of the most puzzling features of science generally, and quantum mechanics specifically. I am not a physicist and cannot say how all this will be received by physicists, but, for myself, it has provided a more profound grasp of quantum theory and left me more hopeful about the possibility of arriving at better answers than we now have. At a personal level, it has given me much food for thought, about both dreams and paranormal (psi) phenomena.

There are two converging lines of thought that are helpful in this respect. The first is, as already noted, the author's approach to an epistemology that unifies subject and object, mind and matter, in a way that does justice to both. The second is his adaptation of David Bohm's approach to wholeness. The latter rests on the postulate of an implicate order as the ground of connectedness and an explicate order comprising the seemingly discrete objects of the world about us and including ourselves. It is the relationship of these orders that has to be understood if the reality of interconnectedness is ever to become the focal point of our existence.

In regard to the first of these, his epistemology of wholeness, Dr. Rosen embarks on a search for a language that maintains the intimate connection of subject and object. He sees poetry as doing just that. The expressive communicative quality of a poem is an embodiment of the poet's own self. Ordinary discursive language maintains the split. We stand apart and talk about things. We need a prose that, in the author's words, maintains the "poetic vision." Isn't the imagery of our dreams a visual analogue of a poetic language, a language that captures the felt reality of the dreamer in the form of the communication itself? True, there is an important difference. The poet sends her poem out into the world. The dreamer is communicating to himself. Both use the medium of metaphor, the poet by rearranging words, the dreamer by rearranging images. Both convey felt depth and embodiment of content far more effectively than ordinary language. Both convey felt truths, whereas ordinary language lends itself to concealment as readily as to revelation of truth. In their content, both the dream and the poem are concerned with connectedness, the dream to deeper aspects of the self, the poem to the deeper reality of the self-in-the-world. Both, in their concern with connectedness, reach deeper into the implicate order, the ground of all being. Both, in their reliance on metaphor, defy objectification. Both reverberate beyond any theoretical interpretive box built to accommodate them. In a society where personal and social truths are so often carefully concealed, perhaps dreams and poetry should have a greater priority than is the case now.

The author's way of working through the intimate relations of local and nonlocal effects at the level of mind and matter has particular relevance for parapsychology. Not only has orthodox science sought to sweep psi effects under the rug, but, as Dr. Rosen points out, researchers in the field have to a considerable extent been standing in their own light. Working as "objective" scientists, they have relied on quantitative methods fashioned for detached observation of an external world. Even in recent years, where the "observer effect" has been noted, there has been a failure to focus on the field dynamics that link the observer and the observed. The nonlocal quality of psi effects seems to betray a deeper order of connectedness. Dr. Rosen's theoretical approach to these strange happenings goes further than earlier ideas that sought to account for this connectedness on the basis of dimensional folding. Focusing on the internal dynamics that propel dimensional change, he arrives at a more complex but well-reasoned account of the dialectic leading to higher dimensionality.

Dr. Rosen shares with Dr. Bohm a deep concern with the level of social fragmentation that now exists and threatens to disrupt the fabric of civilization. Influenced by Jung's ideas about the collective aspects of the *psyche*, he goes beyond Jung in his search for a universal archetype, one that would link the body politic to its organismic roots analogous to the way the archetypes, rooted in the collective unconscious, serve a chthonic function for the individual. He urges us on to view both individual and social pathology in a larger frame of "epistemopathology" and "epistemotherapy."

Although the various chapters were written at different times, the temporal sequences are skillfully woven together. Aside from the scope and originality of the subject matter, there is a certain elegance to his style that succeeds in maintaining a delightful element of suspense. Is he going to succeed in this ambitious undertaking? I think he already has, both in the newer scientific and philosophical perspectives he has set forth and in the manner in which he has addressed the transcultural task of building a bridge between very disparate worlds. Characteristically, in his ending he points the way to a new beginning as he outlines the tasks that lie ahead. I look forward to more to come.

Montague Ullman, October 1992

PREFACE

In his celebrated book *Wholeness and the Implicate Order*, physicist/philosopher David Bohm (1980) pointed to the widespread fragmentation besetting humankind at this critical juncture in our history. Chronicling the pervasive disorder evident on every level of human affairs—personal, social, national and international, planetary—Bohm voiced the urgent need to attain *wholeness*, a term he noted is consanguineous with *health* or *healing* (Bohm 1980, p. 3).

Around twenty years ago, I joined the swelling ranks of those involved in the quest for wholeness. Like Bohm's, my own efforts have been directed toward addressing fragmentation in the sphere of intellect and thinking. *Science, Paradox, and the Moebius Principle* is the culmination of that work—a collection of my philosophical essays published over the past two decades. These papers confront basic anomalies in the foundations of contemporary knowledge, paradoxes that call into question our conventional thinking about space, time, and the nature of human experience. Although the essays range over a variety of disciplines—theoretical physics, mathematics, biology, Western and Eastern philosophy, psychology, and parapsychology—they are woven together by a common thread: the *Moebius principle*. This concept of dynamic process and self-consistent wholeness grew out of my intensive phenomenological investigation of the Moebius surface and several other enigmatic structures found in qualitative mathematics and graphic art. By applying the Moebius principle to diverse manifestations of crisis in modern thought, I have sought to provide an integrative insight that might contribute toward surmounting our dilemma.

In this volume, my previous writings are organized into three main sections. The essays of part 1 apply the Moebius principle to problems in theoretical science (chapters 2–5) and to more general philosophical matters (chapters 1 and 6). In part 2 (chapters 7–11), attention centers on a philosophical examination of the controversial field of parapsychology, the study of anomalous phenomena such as extrasensory perception and "mind over matter" (it is in this area that Montague Ullman, author of the foreword, has made significant contributions; e.g., see Ullman, Krippner, and Vaughan, 1989). Part 3 is devoted to my dialogue with David Bohm, the physicist and philosopher whose thinking has had a substantial impact in areas of primary concern to the book. Included in this section are my two essays on Bohm's work (chapters 12 and 14) and my personal correspondence with him (chapter13). The volume concludes with an epilogue that offers a capsule summary of the Moebius principle, along with a brief indication of the current direction of my work.

Within each of the main parts of the collection, the essays are presented in their original chronological sequence. This will give the reader a sense of the way in which the Moebius principle has evolved, from its inception to its most recent expression (which is by no means its *final* expression, as the epilogue makes clear). No major changes have been made in the core substance of the essays. Revisions reflect only the intention of improving clarity, eliminating technical errors and excessive repetition, and establishing stylistic consistency for the book. However, on the question of repetition, I must call attention to a special editorial problem I encountered and the procedure I have used to address it.

When the works of a given thinker are related to one another through one or more general themes or abstract principles, the particular works themselves can vary substantially in their specific content. In each separate piece, the principle can be restated in a rather distinctive way, so that, were these papers to be assembled as the chapters of a single volume, repetition from chapter to chapter could be minimal. But the principle that serves to integrate my own past works is not just an abstract generality; the Moebius principle is a highly specific model with a set of definite characteristics. As a consequence, on each separate occasion on which I applied this idea to a given field of knowledge, it was necessary for me to repeat my description of it in rather similar terms. The editorial ramification of this for the present volume is that I have had to deal with an unusual amount of repetition. How could I handle this difficulty while maintaining the continuity and coherence of the works, as well as their original substance? Where required, I have omitted whole sections and have bridged the resulting gaps by connective text appearing in italics enclosed by square brackets and written in the present editorial voice. In addition, I have inserted new *endnotes* in certain places, using the same typographic convention of bracketed italics; in the body of the text, the numeric superscripts referring to these notes also are bracketed and thereby are distinguished from original superscripts.

Since this collection is interdisciplinary in nature and largely nontechnical, I believe it should appeal to philosophically oriented readers with highly diverse academic backgrounds. For those who sense the need for fundamental change and are aware that issues in the foundations of science, philosophy, and parapsychology may be critically relevant to this, the book should be of special interest.

The appearance of *Science, Paradox, and the Moebius Principle* seems well timed. Beginning with Bohm's book on wholeness mentioned above, the past decade has seen the publication of a number of volumes focused on the paradoxes of modern science, the crisis in contemporary thought, and the necessity of a basic new insight to bring order and wholeness from the current state of fragmentation. SUNY Press alone has published several titles along these lines (e.g., Comfort 1984; Griffin 1986a, 1988). I believe my own contribution is uniquely responsive to the challenge posed in at least three important respects.

The *Moebius principle* constitutes a process theory and philosophy that is *thoroughly* processual. In this approach, the elemental unit of process is *paradox*, the dynamic tension of juxtaposed opposites. But paradox does not simply go unresolved (as in purely relativistic accounts), nor is it merely averted or eliminated by extrinsic means (as in various forms of absolutism). Instead, an *intrinsic* resolution of paradox is provided which results in the *re-emergence* of paradox at another, more complex level. So, in effect, an idea of change is advanced in which process itself is "processed." This notion is explicitly articulated in part 1 of the book, where the "self-processing" of space and time is brought to light. That is, when the *Moebius principle* is applied to the foundations of biology and physics, a novel concept of 'dimensional generation' appears in which space and time themselves, animated by the inner workings of paradox, undergo continuous internal transformation and organic growth (see especially chapters 4 and 5).

A second original feature of the *Moebius principle* is that it furnishes a *highly specific* account of dialectical change. The geometric forms that are employed help to concretize the dynamics of paradox, to bring them into focus with unusual clarity. In the course of this, important details of process are disclosed that remain unsuspected in more nebulous approaches.

Finally, I note the kinship of my work (especially in its later development) to modern phenomenological philosophy. To be sure, thinkers like Heidegger (1977a)[1] and Merleau-Ponty (1968) tellingly have questioned the presuppositions of Western thought. Moreover, this line of questioning has been carried further than merely programmatic and general philosophical critiques. It has been brought to bear on specific theoretical structures encountered in the foundations of science and mathematics, leading to the implication of alternative approaches. For example, Heidegger offered a probing analysis of the classical concept of *space*, characterizing it essentially as sheer externality, as the " 'outside-of-one-another' of the multiplicity of points" (1962, p. 481). Heideggerian reflection discloses the possibility of a very different intuition of space, one in which the elements of space are apprehended as overlapping or interpenetrating, as being *internalny* related (see chapter 5). This indeed has important implications for addressing basic theoretical problems in science and mathematics. Yet by and large, the Heideggerian alternative has not taken hold, and I suggest this is essentially because it originates *outside* the domain it challenges. In Heidegger's deconstruction of space and of other scientific concepts, science implicitly is taken as *object*, in the sense that the language system and frame of reference used are not those of science itself; the problems posed are not science's own problems as *science* would define them.

In the present book, a *self*-deconstruction of science evolves. Questioning the foundations of contemporary scientific thought from *within*, exploring paradox from science's own perspective as it arises at the *inner* boundary of science, *Science, Paradox, and the Moebius Principle* aims to open a natural pathway *beyond* the prevailing framework of knowledge and experience.[2]

Nevertheless, an "internal deconstruction" of this kind creates a unique problem in communication. Were my mode of discourse either clearly outside or simply inside the language system of science, my readership would be readily definable. In the former case, the appropriate audience would be unequivocally nonscientific, with working scientists having little serious interest in this book (at least in their capacity *as* scientists). Conversely, in the latter case, it would be the scientist to whom primarily I would be speaking, and I would limit myself to his or her technical, relatively narrow frame of reference and tacit set of rules for how one speaks; from such discourse the nonscientific reader would be excluded completely. (Indeed, this tacit linguistic barrier is the way science long has protected itself from effective challenge.) But in the "deconstruction" I actually am attempting, no such "monaural" channel of communication is possible. Inasmuch as I assume the perspective of science, the reader with no ear at all for science will find me incomprehensible. And since I raise fundamental questions about science carrying me *beyond* its traditional purview, the reader receptive only to the style of speaking that has been strictly prescribed by science is bound to mishear me. I suppose I am requiring of my audience a kind of "stereophonic listening." Or let me put it this way: The success of this project depends not merely on the receptiveness of transdisciplinary scientific readers but on the *"transcultural"* reader, in the general sense of C. P. Snow's "two cultures" of science and the humanities. Although the background and training of such readers may bring them to science from the outside, they will be predisposed to extending themselves across the "cultural" divide. By the same token, "transcultural" readers initially socialized *within* the language system of science will not be dismayed by the "foreign dialect" they may hear but will be inclined to listen in a new way.

Over the years, a great many people have contributed to making this book possible. Here I will limit myself to expressing special appreciation to just four individuals. First, I must acknowledge that the impetus for this project came from psychologist/musician Joel Funk, who kept urging me to put out a "recording" of "Steve Rosen's Greatest Hits." More immediate encouragement has come from Sal Restivo, the SUNY editor who prevailed on me to see the project through at a time when I was ready to shelve it (Sal's own deconstruction of science is well known; see, for example, Restivo 1988). It is my sister, Myra Schwartz, who originally brought my attention to the Moebius paradox, and since that time in 1972, we have spent innumerable hours fruitfully discussing it. Last and surely not least, I give my heartfelt thanks to my colleague and life partner, Marlene Schiwy, for her many perceptive editorial suggestions and for the understanding and loving support she has given me, from the start of this project to its finish.

New York, April 15, 1992

I would like to add special acknowledgment to one other person. As I noted above, part 3 of this volume is devoted to my interactions with physicist and philosopher David Bohm. No thinker has had a greater impact on the substance of the book.

Today, as I was on my way to mail a letter to David, I was shocked and saddened to learn from a colleague that he had just passed away. His leadership in our quest for wholeness will be greatly missed.

New York, October 29, 1992

PART I.

The Moebius Principle in Science and Philosophy

INTRODUCTION

The papers in part 1 span a seventeen year period (1975–1992). The section begins and ends with essays written in a broad philosophical vein. In chapter 1, the Moebius principle is set forth for the first time and applied to such questions as the nature of dialectical change, the meaning of relativity, and the issue of human freedom. Note that this chapter also gives voice to two of the most prominent concerns of part 1: the issues of symmetry and paradox. Chapter 6 closes this first part of the book by addressing the timeless philosophical question of mind, matter, and their interrelationship.

In the intervening chapters of part 1 (chapters 2–5), the Moebius principle is specifically brought to bear on some challenging problems of theoretical science. The procedure followed in each case is to raise a fundamental difficulty within a given province of science, then demonstrate that a new insight can be achieved, a genuine solution approached, by applying the Moebius principle (in chapter 4, the application is spelled out less explicitly than in the other chapters of this group).

In chapter 2, the issue of symmetry is made the primary focus, especially as it relates to the matter of parity violation in modern physics. Next, in chapter 3, the question of biological evolution is explored, and four levels of nature are identified and discussed. With chapter 4, attention turns to the role of infinity in the quantum and relativity theories. Then the Moebius principle reaches its most mature expression in a chapter dealing with the problem of unifying the forces of nature. Note that chapter 5 (and to a somewhat lesser extent, chapter 4) may prove particularly challenging for readers with no prior knowledge of conceptual issues in mathematical physics. However, though such readers can choose to bypass this material without jeopardizing their overall understanding of the book, I believe that all "transcultural" readers will be able at least to glean the gist of chapter

1

5, and I would urge them to make the effort, since it does contain the most detailed, refined, and ambitious statement of the Moebius principle thus far in print.

1

The Unity of Being and Becoming (1975)

It appears that we soon will witness the emergence of a new image of world and man.[1] I believe the scope of the change should be virtually unbounded, encompassing long-held and deep-rooted views of reality. Moreover, I suggest that the impending transformation reflects the workings of a process that is essentially dialectical. [concerned with or arising through opposing forces]

Students of philosophy are familiar with Hegel's famous triad: thesis → antithesis → synthesis. This is the dialectic. For every idea, there exists or develops a contradiction. Change occurs through the reconciliation of opposites. Thesis is not merely replaced by antithesis; rather, opposites are *fused*, canceling the exclusive validity of either. A broader domain arises wherein old "opponents" may be relegated to the status of "special cases," if they are not discredited completely. Subsequently, the synthesis itself may be recognized as a higher-order thesis with its own developing antithesis. The circle closed by synthesis is thus reopened, the dialectical process taking on the general appearance of a mounting spiral.

A mounting spiral

The metamorphosis I am anticipating may be understood in these dialectical terms, seen as constituting a novel synthesis. In fact, the new worldview should encompass an opposition that is quite fundamental—that expressed in the notions of *Being* and *Becoming*. Being is associated with changelessness, Becoming with change. If the reconciliation of these conflicting themes indeed is in the works, is there a concrete image capable of portraying their integration in a graphic way? Can we develop a specific model of the synthesis that could serve as a guide in our time of transition?

THE WORLD AS BEING/BECOMING

Three Levels of Changelessness

To the naive observer, changelessness is equated with absolute rest. The ancient Greek philosophers of Elea (among whom

Parmenides was the leading figure) promoted the doctrine of Being at this most obvious level. Zeno, the chief student of Parmenides, is noted for his paradoxes of motion, one of the most famous of which involves Achilles's futile effort to overtake a tortoise in a foot race. This is the best known example of the Eleatic attempt to prove the impossibility of motion.

Democritus, another great philosopher of that era, ostensibly reinstated the idea of change. Instead of supposing that space was full (an underlying presumption of Eleatic philosophy), he assumed the existence of a void through which bodies could move freely. However, the large-scale, perceptible motions that seemingly brought about change were reducible to the imperceptible movements of atoms, which were thought to be themselves immutable. Since atoms were the unchanging bedrock of reality, the transformations that took place in the large-scale world were considered to be without real substance. Thus Democritus's viewpoint actually affirmed the Parmenidean doctrine of an "unchanging reality. . . behind or beyond the world of [changing] appearances" (Kerferd 1971, p. 34).

Philosopher Milic Capek (1961) informs us that all classical thinkers, including the atomists, had difficulty in accepting the primary quality of motion: "atomism, ancient as well as modern (Newtonian physics, for instance), retained the basic Eleatic belief in the constancy of being which in its quantity as well as quality is not subject to any change; change, though not completely denied, retained a half-spurious character, as it did not affect the ultimate units of material substances" (Capek 1961, p. 137).

This classical paradigm of Being, which in one form or another prevailed for centuries in the West and dominated the science of the nineteenth century, was significantly challenged by Einstein's theory of relativity. But was the new paradigm one of Becoming (change), or was changelessness being espoused at still another level?

At first glance, relativity—especially Einstein's *general* principle of relativity—does seem to transform the classical concept of Being into one of Becoming. In the classical view, bodies remain essentially the same regardless of their apparent motions, which are presumably instigated by outside forces. But a fundamental proposition of relativity theory is that force is unified with that upon which force acts; that is, force is *internalized*. The gravitational attraction of bodies is geometrically interpreted in the general theory as a curvature of the space-time continuum. Yet space-time curvature does not express attraction as abstracted from the bodies themselves; rather, the bodies are identified with the curvature. The persistence of classical thinking might lead one to conclude that without the curvature, bodies would not be attracted to each other. The more appropriate statement is that without the curvature, there are no bodies! Hence, bodies are not static beings, for interactions are internal to their essences. Bodies *are* interactions, matter is energy, Being is Becoming.

Or is it? Is Being transformed into Becoming by this dynamization of bodies? It depends on what is demanded of the concept of Becoming. In the present context, our demands are maximal. Becoming must entail change in its deepest sense to be a suitably potent antithesis of Being. Change has to involve genuine novelty, an irretrievable break with the past. We should not be able to negate it by reversal and so recover the state of affairs existing previously. In Capek's words, "novelty implies irreversibility" (Capek 1961, p. 341). However, Einsteinian relativity appears to mandate reversal in the end, for Einstein sought a closed four-dimensional geometry that would serve the classical aim of completely encompassing all the phenomena of the world. Such a self-contained, ultimately self-reflexive geometry would permit no further change. Indeed, Einstein himself came to the conclusion that three-dimensional Becoming may be conceived as Being in the four-dimensional realm (see Capek 1961, p. 158)— that what appears to us as change is, in a higher sense, unchanging.

In sum, the idea of Becoming requires that motion or change be real, that it be internal to the entities moved (not merely applied to them from without), and that it be irreversible, introducing an entirely new state of affairs, one not permitting recovery of the old.

Being/Becoming: The Incubating Synthesis

Philosopher Nicholas Georgescu-Roegen (1971), a strong proponent of the dialectical approach, recently suggested a means of integrating the antithetical concepts of "reversibility" and "irreversibility." In the standard definition of these terms, a reversible process entails a complete return to an earlier state or

condition, whereas an irreversible operation involves an irretrievable departure. Deviating from conventional nomenclature,
Georgescu-Roegen renamed the latter an *irrevocable* process,
reserving the term *irreversibility* for his integrative concept. The
"irreversible" event, in Georgescu-Roegen's sense, proceeds in one
direction but returns periodically to a previous phase. He offered
the example of a tree growing progressively despite losing its
leaves each year. The general idea is illustrated in figure 1.1
(which is to be interpreted as portraying *temporal* relations, not
merely spatial ones). In a world governed by irreversibility,[2] a
system neither reverts completely to its initial condition (represented in fig. 1.1a by the lone circle), nor are its successive states
entirely disconnected from each other (the irrevocability symbolized by fig. 1.1b). Rather, the states of a system overlap one
another in a spiral fashion (as in fig. 1.1c). Accordingly, Georgescu-
Roegen described the irreversible process as "a series of overlapping irrevocable processes" (1971, p. 200).

Figure 1.1

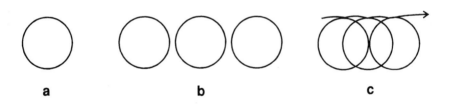

a   b   c

Reversibility (a), irrevocability (b), and irreversibility (c)

Irreversibility, applied to the question of determinism, leads
directly to the notion of *partial* determinism. Thus Georgescu-
Roegen concluded that although the present moment is determined by the past (the influence of reversibility), it must also
possess an element of novelty (the influence of irrevocability).
Capek (1961) brought out the epistemological consequences of
this, asserting that while our knowledge of future states is necessarily incomplete, we successfully can make general predictions.
In his discussion of perceptual models of physical reality, Capek
provided a familiar example that can be interpreted as an instance

of partial determinism. He described an unfolding musical phrase: "The individual tones are not externally related units of which the melody is additively built, neither is their individuality . . . dissolved in the musical whole. . . . Like every dynamic whole, it [the music] exhibits a synthesis of unity and multiplicity, of continuity and discontinuity" (Capek 1961, pp. 371–72).—Of determinism and indeterminism, we might add, based on our phenomenological experience with music. Insofar as the individual tones and tonal combinations persist, are not dissolved as the composition progresses, there is a basis for confirming our expectations. But the elements do not *wholly* retain their distinctiveness. There *is* a certain loss of individuality resulting from the transmutation of tones in the creatively unfolding temporal gestalt, and this is the aspect of our experience with music that cannot be predicted in advance.

Of course, the syntheses of unity and multiplicity, continuity and discontinuity, determinism and indeterminism, are but alternative expressions of the emerging concept we are presently calling "Being/Becoming." Yet Capek spoke of the great difficulty of enunciating "this paradoxical 'unity of opposites'" (1961, p. 372). He himself ruled out the possibility of constructing a visual model able to capture the fusion of Being and Becoming. Though the spiral has been mentioned, it is at best only a vaguely suggestive metaphor. I have asked myself, therefore, if it is possible to embody the union of opposites in a visual image that can be grasped at a glance, as it were, thereby surmounting verbal limitations in conveying the nature of paradox.

The Surface of Moebius: An Image of the Dialectic

In seeking a concrete embodiment of Being/Becoming, let us turn to the qualitative field of mathematics known as topology, and to the *Moebius surface*, a topological curiosity with unique transformational properties. The exercise in visualizing the Being/Becoming paradox begins with the construction of a Moebius strip (fig. 1.2b) by joining the ends of a narrow band of paper in such a way that one of the ends is given a half-twist (rotated through an angle of 180°). Next take another strip of paper and join the ends without twisting; this will produce a simple ring or cylinder (fig. 1.2a). Now, on one side of each of these strips place an asymmetric profile, like that shown in figure 1.2. (It does not

Figure 1.2

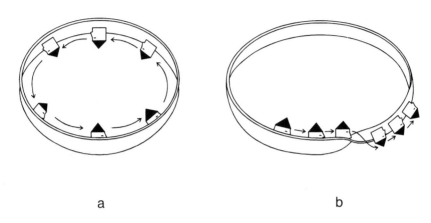

a   b

Cylindrical ring (a) and Moebius strip (b)

matter whether the profile faces left or right. What is critical is
that it be asymmetric, for if a symmetric figure were chosen, an
important difference between the two surfaces could not be
observed.)

If we now imagine that this profile can move along the
surface of either strip (as indicated by the arrows in fig. 1.2),
we see that in the case of the cylindrical ring, the profile can go
around and around indefinitely on the inside without ever coming
into contact with the outside. (Crossing over the edge of the paper
would break the continuity of movement and is thus prohibited
in this exercise.) Here the cylindrical ring is seen to possess two
independent, disconnected sides and is classified by topologists
as a two-sided surface.

In the course of moving the profile along the Moebius strip,
we discover its unique feature: the profile is transported from one
side of the strip to the other without crossing an edge. This illus-
trates the mathematical attribute of *one-sidedness*. In both the
cylindrical ring and the Moebius strip, we can match a point on
one side with a corresponding point on the other. But in the
former case, this pairing is superficial in a mathematical sense.
You must do the matching, for the fusion of points is not inherent
to the geometry of the ring. Consequently, points may be regarded

as insulated from and external to one another. It is true that in the Moebius case, if you place your index finger on any point along the strip, you will be able to put your thumb on a corresponding point on the opposite side. The Moebius strip does seem to have two sides, like the cylinder. But this holds only for the cross-section of the strip defined by thumb and forefinger. Taking the full length of the strip into account, we find that points on opposite sides are intimately connected—they can be thought of as "twisting" or "dissolving" into each other, as being bound together *internally*. Hence the topologist defines such pairs of points as *single* points, and the two sides of the Moebius as but *one* side. Already we begin to see in the Moebius surface a visual/geometric representation of the union of opposites.

The ability of the Moebius strip to serve in this capacity extends still further. There is a sense in which it can accomplish a *restoration of symmetry.*

Symmetry may be defined in many ways. For our purposes, let us define it in terms of mirror reflection (which involves a correspondence in size, shape, and relative position of parts across a dividing line). The full-faced, two-dimensional form of figure 1.3 is judged symmetric with respect to the plane of the page because it corresponds exactly with, and thus can be carried into, its mirror image without being removed from that plane. On the other hand, the two-dimensional profile cannot be superimposed on its mirror image in the plane and is therefore considered asymmetric with regard to the plane.

But the profiles can, of course, be brought into point-to-point correspondence if one is lifted from the plane through the third dimension and turned over onto the other. Since the rotation necessary to superimpose right and left profiles is thus perform-able, the asymmetry of the profile may be considered relative. Though the profile is asymmetric with respect to the plane, a hyperdimensional rotation out of the plane can restore symmetry, in the sense that the necessary matching can be achieved.

Can the symmetry of the right-facing profile revolving around the cylindrical band be restored? To turn right into left, all we would need to do is turn the profile around in its place, rotate it through an angle of 180° about its own axis. The trans-formation of right into left also could be accomplished by trans-porting the profile over an edge of the band to the other side. But in neither case would the operation be inherent to the topology of the two-sided cylinder.

Figure 1.3

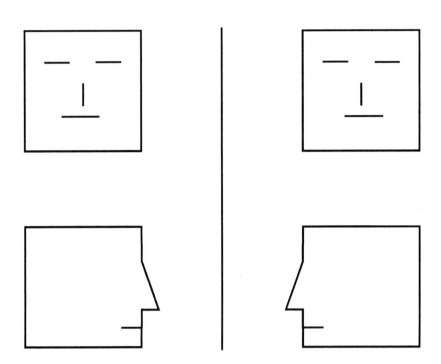

Mirror reflection of symmetric (upper row) and
asymmetric (lower row) two-dimensional figures

The structure of the cylinder can be clarified by imagining
a plane of symmetry bisecting it along its transverse (long) axis
(here you might picture a large square of cardboard thrust
through the paper ring, partitioning it lengthwise into two
identical narrower rings). The profile of figure 1.2a moves in
conformance with cylindrical structure if and only if its rotation
is confined to the cylinder's plane of symmetry. This requirement
clearly is not met by either of the above-mentioned operations
designed to restore the *profile's* symmetry. The contrasting
structure of the Moebius surface is revealed in the fact that it
possesses no plane of symmetry like that of the cylinder. Thus,
rather than restricting the profile to the two-dimensional plane
that maintains its asymmetric orientation, rotation consistent

with the Moebius structure carries the profile into the third dimension, where its orientation is reversed (as shown in fig. 1.2b). So we see that on the Moebius surface, right is transformed into left geodesically, that is, by following the natural contour of the surface. Here, symmetry restoration *is* an internal feature of the topology.

The symmetry-restoring property of the Moebius strip confirms it as a convincing embodiment of dialectical process. The union of opposites can now be understood as the restoration of symmetry (fusion of right and left) by hyperdimensional rotation, which is uniquely accomplished by the Moebius. Yet is symmetry *restoration* enough? The interchange of Being and Becoming has been conceived as an opening spiral. The synthesis that resolves the conflict becomes a higher-order thesis calling forth opposition in a new arena. Every closing circle must point beyond its boundaries to novel potentialities for action, lest Being prevail over Becoming. This is the requirement of irreversibility to be met in the present context by denying the finality of symmetry restoration. The surface of Moebius actually satisfies this condition, for it not only integrates opposites, reconciling mutual asymmetry;[3] it also is *itself* asymmetric (as already noted above, in observing that, unlike the cylindrical ring, it possesses no plane of symmetry). When the Moebius surface is formed, the paper can be twisted one way to produce a right-facing configuration or the other to create its mirror-reversed counterpart.

We next encounter a higher-order topological structure, the *Klein bottle* (named after its discoverer, the German mathematician Felix Klein). The Klein bottle superimposes the right-twisted Moebius strip on its left-twisted mirror image, just as the strip superimposed the opposite-facing profiles. However, whereas the strip requires three dimensions to perform its function, the Klein bottle entails the fourth. This is why Klein-bottle operations can be conceived by us three-dimensional creatures only by exercising topological imagination—construction of the Klein bottle cannot be executed in three-dimensional space. When we employ such imagination, we realize that the Klein bottle, like the Moebius strip, must be asymmetric. If there is a right-oriented Klein bottle, there must be a left-oriented counterpart; and the existence of these opposites suggests a still higher topological form, the "hyper-Klein bottle," which would require a fifth dimension to accomplish its fusion of opposites. Clearly, the process being described is endless. We may say, in general, that the completion of hyperdimensional rotation at one level signals a commencement of rotation at the next. The emergence of novelty proceeds in this manner by hyperdimensional ascension. This is the dialectical spiral.[4]

HUMAN BEING/BECOMING: A QUESTION OF FREEDOM

In the Moebius-Klein generality, the union of Being and Becoming is seen in its broadest scope. Here is a process that ultimately supersedes the limitations of any particular domain of space and time, since, indeed, it is the operation by which new domains are ceaselessly created. If this is the way reality unfolds—as a spiral of general reoccurrence and progressive advance that incorporates genuine novelty—should we not expect the same pattern to apply to the dynamics of a *given* space-time realm? In fact, the notion of partial determinism discussed above, wherein the future is neither strictly determined nor wholly indeterminate, evidently is just such an application. The example of a developing musical phrase dealt with our own particular dimension of time—our *fourth* dimension.

The question now is, can we, as human beings, play an active, deliberate role in shaping our dialectical advance, or must we conform to the pattern passively, being limited to the role of spectators? In other words, is the indeterminateness of our world to be understood as sheer randomness, or can it incorporate an element of freedom?

Surely we are not free-floating spirits detached from the worldly continuum. None of us is a *deus ex machina*, able to swoop in and effect change arbitrarily, in whatever manner we wish. Hence we cannot act against the grain of physical reality; in this sense, we are not free. But the human individual is not simply a *part* of the world. It seems more appropriate for us to think of ourselves as *aspects*. How do these two conditions differ? In attempting to distinguish the meaning of an aspect, let us experiment once again with a visual cue, this time with a figure from Gestalt psychology, the *Necker cube*.

The Necker cube (fig. 1.4a) is a two-dimensional projection into the third dimension. When you look at this structure from one perspective, it appears to hover above your line of sight. Then,

Figure 1.4

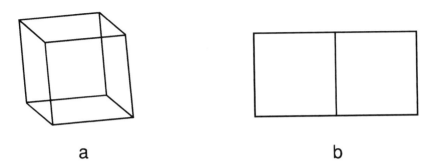

Necker cube (a) and divided rectangle (b)

suddenly shifting to the other viewpoint, you see it as if it were below you, and all the faces that seemed to be outside presently appear to lie inside. Though this reversal shows that the cube can be seen from two distinct perspectives, both are *aspects* of the cube, because each encompasses the whole configuration. Neither perspective should be thought of as merely contained *in* the cube the way the squares of figure 1.4b are contained within the divided rectangle.

Returning to the Moebius strip, its one-sided character implies this quality of aspecthood. At any and every point along the surface, we must concede that a given side is but one of two sides. Yet when the entire length of the strip is considered, each side covers *both* sides, due to the intimate connectedness of the one-sided surface, just as each perspective of the Necker cube, being an aspect, uses the whole to express itself.

Before examining the broader significance of these observations, I would like to underscore the paradoxically reflexive nature of Moebius movement. The totality of the Moebius strip is embraced by each of its sides owing to the fact that the sides are internally bound together as one. Therefore, since sides share a common identity, Moebius passage through the higher dimension that carries one side into another geodesically may be said to involve the turning back of the side upon *itself*. Of course, the distinctness of the sides is not simply negated when the Moebius structure is taken as a whole; the Moebius surface is not "one-sided" in the simple, undifferentiated sense of the single side of the cylindrical ring. The paradox is that, although opposing sides of the Moebius are fundamentally identified, they maintain their *difference* as well; the Moebius is one-sided while *also* being two-sided (the standard topological classification only names the former). Thus, in the Moebius transformation, reflexive self-reference and reference to other are thoroughly blended. Enveloping the whole, the Moebius aspect turns back upon itself and, at the same time, upholds what is different.

It may seem difficult to extend the relation of the aspect to the whole found in the Necker cube and Moebius strip to a person's relation with the world around him. In this regard, it is interesting to note that physicist George Gamow's (1961) discussion of the "unusual" properties of the Moebius strip appears in the same chapter wherein he describes a mental/topological operation through which a man transforms himself so that the entire universe is squeezed into the cavity of his body (see fig. 1.5). The objective world is entirely encompassed by one of its subjects in this "inside-out" universe. Here, the subject is not just a part of the world; by virtue of his ability to embrace it, he is an *aspect*.

In a similar vein, philosopher Oliver Reiser speculated that "the atom is a galaxy turned inside out, rotated through a higher dimension" (1966, p. 412). But while Reiser emphasized the general importance of "circumversion" (the action of turning inside out [Reiser 1966, p. 495]) and stated that it is only possible by a hyperdimensional movement, he was uncertain as to just how the performance of such an operation might be conceived. The topology of the Moebius surface offers a clue in this regard; it provides a basis for the occurrence of circumversion through a natural, internal transformation.

We see more clearly the implications of "hyperdimensional aspecthood" for our issue of human freedom when we recognize that the human is the being who projects himself into the fourth dimension—"the being who hurls himself toward a future and who is conscious of imagining himself as being in the future" (Sartre 1957, p. 16). It is in this reflexive action, where subject becomes aware of itself as object, that man is an aspect of the world, not just a part. Identified with the world yet distinct from it, man can incorporate the world without losing his individual identity in the process. Though he is only a "side" of reality, a

Figure 1.5

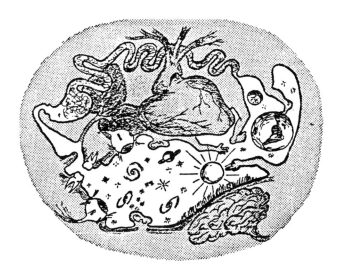

Inside-out Universe
From George Gamow's *One Two Three. . . Infinity*
(New York: Bantam Books, 1961).

side that is an aspect may spread its influence to the whole in a manner impossible for an isolated part. In the subject's intimate connection with the objective world—in our ability to propel ourselves into its future, its higher dimension of potentiality— each of us can consider possibilities and choose a course of action. What is indeterminate, we can determine. As aspects of the world, we are free.

2

Synsymmetry (1975)

The violation of parity in weak interactions demonstrated in 1956 was an event that shook the foundations of physics. Since that time, the status of symmetry has been very much in doubt.

The problem is presently addressed by first reexamining the essential relation between symmetry and asymmetry. Then, through the intuitive medium of visual geometry, an attempt is made to show how these opposites may be fused in a topological structure expressing a new principle, that of *synsymmetry*.

SYMMETRY

A body or spatial configuration satisfies the criterion for symmetry if it can be superimposed on its mirror image within a given dimensional frame of reference *[as defined in chapter 1]*. For example, a frontally oriented, two-dimensional figure *[fig. 1.3, top]* is judged symmetric with respect to the plane of the paper because it can be carried into its reflection without being removed from that plane. On the other hand, a two-dimensional profile *[fig. 1.3, bottom]* cannot be superimposed on its mirror image unless it is lifted from the plane and properly turned in the higher (third) dimension. Therefore, the profile is deemed asymmetric. Of course, we *can* perform the rotation necessary to superimpose right and left profiles. From this it is clear that the profile is asymmetric only with respect to the plane. A hyperdimensional rotation out of the plane of the paper achieving the point-for-point alignment of profiles may be said to "restore" symmetry.

What about asymmetric *three*-dimensional forms? Apparently, no fourth spatial dimension is known through which a solid asymmetric figure can be transformed into its mirror image. It would seem that the asymmetry of the solid is basic. However, the existence of a real-world equivalent of the solid's mirror image indicates that the *effect* of a hyperdimensional rotation has been achieved. For example, the mirror image of our left hand exists in the real world as our right hand. Since the right parts of our bodies are generally mirror images of those on the left, we are regarded as bilaterally symmetric. The fundamental symmetry of the physical world is assumed by presupposing the existence of real-world counterparts for the mirror images of *all* asymmetric bodies.

Yet an empirical search for real-world replicas of mirror image configurations cannot be conclusive on the matter of asymmetry. Though the complement of the left-twisting amino acid, for instance, is not found in local biological evolution, its appearance elsewhere in space or at another time is not precluded in principle. Unless and until we discover a case of an unreflected, right-twisting amino acid, we can make no decision about symmetry. It follows that, at this level of analysis, we can never establish asymmetry as basic. To arrive at an unequivocal conclusion concerning symmetry, we must operate at the level of the space-time principles themselves. If a mirror event does not violate the presumed structure of space-time composed from the laws of physics, the event that produced the image is symmetric in principle. Inversely, if a mirror event does violate physical law, then it is certain that the event that produced the image is asymmetric with respect to the underlying space-time structure, for the real counterpart can nowhere be found in the universe built from that structure.

DIALECTICAL INTERPLAY OF SYMMETRY AND ASYMMETRY

Mathematician Hermann Weyl introduced his work on symmetry by drawing from the field of art:

> "Symmetry," says Dagobert Frey in an article *On the Problem of Symmetry in Art*, "signifies rest and binding, asymmetry motion and loosening, the one . . . formal rigidity and constraint, the other life, play and freedom." Wherever God or Christ are represented as symbols for everlasting truth or justice they are given in the symmetric frontal view, not in profile. (Weyl 1956, p. 678)

There is no problem of greater significance for comprehending the physical world than that of symmetry. The laws of physics are woven together in a mathematical fabric whose essential properties hinge on the question of symmetry. Thus Weyl wrote that "the mathematical laws governing nature are the origin of symmetry in nature" (1956, p. 674) and that "the entire theory of relativity. . . .is but another aspect of symmetry" (1956, p. 681).

It cannot be denied that both symmetry and asymmetry are manifest in the world. But the structure of reality need not be revealed in its phenomenal appearance. In fact, physical science has traditionally relegated asymmetry to secondary status: every asymmetry was believed to conceal a more basic symmetry.

On the other hand, *biological* evolution may be regarded as a progressive narrowing of symmetry operations (as noted by Weyl). In morphogenesis, for example, living forms have displayed a degeneration of spherical symmetry into a lesser bilateral symmetry. And as concerns function, the loss of reproductive potential (i.e., the capacity for a system to be biologically transformed into itself) with evolution into more complex forms (see Koestler 1964) may rightly be interpreted as a weakening of symmetry. While the increasing asymmetry of molar organic form and function does not necessarily imply any deeper molecular asymmetry, it is known now that the deeper asymmetry does exist: "all subunits of protein and nucleic acids [are] lefthanded," as Gardner (1964, p. 142) succinctly puts it. The discovery of life's primary asymmetry, initiated by the work of Pasteur on racemic acid and more recently highlighted by the efforts of Watson and Crick, broadens the arena of interplay of symmetry and asymmetry and makes it doubtful that this dialectic is purely superficial.

One is reminded of the remarkable conception long ago advanced by Albert Lautman, a French mathematician. Lautman viewed the subjective "world of sensation" and those of chemistry, physics, and mathematics as all participating in "an intimate union of symmetry and dissymmetry" (1971, p. 45). According to Lautman, "a phenomenon can only exist in an environment possessing its characteristic symmetry" (1971, p. 45), but this symmetry must always be of a limited kind. He invoked Pierre Curie's concept of the "limited symmetry" of physical phenomena, defined by "the presence of certain elements of symmetry in conjunction with the necessary absence of other elements" (Lautman 1971, p. 45).

This same dialectical juxtaposition of symmetry and asymmetry is observed in mathematical theories, where what is often found is "a division of mathematical entities into two classes, a rigorously symmetric class comparable to an ambidextrous being and a class that mathematicians call antisymmetric, i.e., a class of objects which change orientation under symmetry, as do the left hand and the right hand" (Lautman 1971, p. 46).

Similarly, in physical theories such as the theory of the electron, it was found necessary through experimentation to endow the electron with "a moment of rotation, or spin, which could at any time take on two opposite values with two different possibilities" (Lautman 1971, p. 47). After further elaborating the relation of the electron theory to mathematics, Lautman stated: "This is a typical case in which the physics of dissymmetric symmetry leads back to an algebra in which opposite terms exchange roles" (Lautman 1971, p. 47).

Next, Lautman attempted to articulate his theme in the most general terms. Two ideas are found in the duality of left and right:

> (1) the division of a complete entity into two distinct parts, at least by inversion of their orientation, and (2) the existence of an involute relation between the two parts, such that A is to B as B is to A, the symmetry reapplied yielding again the original element. . .this situation is equivalent to the possibility of distinguishing within a single entity two distinct entities X and X' which will be said to be in a dual relation, if an orientation or an ordering can be found for each of them such that they are inverse to one another, and if, in addition, we can find an involution relating them, i.e., if X is to X' as X' is to X or (X')' = X. (Lautman 1971, p. 50)

In short, Lautman saw the basic mathematical relation as involving both asymmetry and the restoration of symmetry by involution. Here the left-right distinction is fundamental, and yet, to prevent an absolute asymmetry just as abhorrent as absolute symmetry, the involute operation that reinstates symmetry is always specified.

However, Lautman might have placed greater emphasis on a significant point. The restoration of symmetry cannot be a terminal operation. If involution merely reestablishes symmetry and does nothing more, the dialectical tension of left and right dissolves in a final way, and symmetry is elevated to primary status after all. Of course, this was not Lautman's intention. He stated explicitly that "the symmetry that we shall call symmetric is nothing but a limiting case of an antisymmetric symmetry which remains the general case" (1971, p. 55). And yet, the "general case" is not fully satisfied by the closed circle that

Lautman himself described. We must reinterpret that circle as an opening *spiral*. Somehow, the restitution of symmetry at level *n* must be accompanied by the introduction of novel *asymmetry* at level *n* + 1. Just such a process will be set forth later.

For the present, it is notable that Weyl also called into question the idea of absolute symmetry. To understand Weyl's crucial discussion, let us consider the basic geometric concept of *spatial transformation*. A transformation of space is a mapping or projection of space such that a point, *p*, becomes associated with another point, *p'*. We focus our attention on a type of transformation called an "automorphism." The automorphism is a projection of points, *p* → *p'*, and the inverse, *p'* → *p*, that does not alter the structure of space. An automorphism may be a simple motion or it may be a reflection. In the former case, the orientation (left or right) of any configuration of points will remain unchanged, whereas automorphic reflection will transform left into right and vice versa. The key question is whether the difference between left and right is fundamental or only arbitrary. If it is arbitrary, if symmetry prevails, if "parity" is conserved (as a physicist would say), then a change in orientation would not involve a change in the underlying structure of space. "What we mean then by stating that left and right are of the same essence is the fact that *reflection in a plane is an automorphism*" (Weyl 1956, p. 682).

Though Weyl stated that "the inner structure of space does not permit us, except by arbitrary choice, to distinguish . . . left from . . . right" (1956, p. 681), he added these important words of qualification:

> If nature were all lawfulness then every phenomenon would share the full symmetry of the universal laws of nature as formulated by the theory of relativity. The mere fact that this is not so proves that *contingency* is an essential feature of the world . . . The truth as we see it today is this: the laws of nature do not determine uniquely the one world that actually exists, not even if one concedes that two worlds arising from each other by an automorphic transformation, i.e., by a transformation which preserves the universal laws of nature, are to be considered the same world. (1956, p. 687)

The problem posed by Weyl is essentially similar to that identified in our discussion of Lautman. Lautman's X → X',

X′ → X relation is closely analogous to Weyl's $p \to p'$, $p' \to p$ automorphism. If we are to go beyond the simply symmetric outlook as both Lautman and Weyl suggested, we must develop a space-time conception that transcends the closed automorphism, one that opens perpetually as a spiral.

Prior to 1956, however, the ideas of Lautman and Weyl were largely disregarded. There was no compelling reason to doubt that the world is built from a wholly symmetric blueprint. Even the asymmetry of life discovered by Pasteur could be accounted for. Physicists had only to assume that the unidirectional twist of organic matter is merely adventitious, a quirk of local evolution whose mirror image appears in real form at some intersection of space-time coordinates presently inaccessible to human observation. I pointed out earlier that fundamental asymmetry would be established if a mirror event were to be produced that would violate physical law. But the deep-rooted bias favoring symmetry led physicists to judge that such an occurrence was unthinkable. Then, in 1956, in the laboratory of Wu and Ambler (Wu et al. 1957), the "unthinkable" came to pass.

THE VIOLATION OF PARITY

The concept of 'parity' is none other than that of mirror symmetry applied to the domain of physics. Parity is conserved if a pattern of physical interaction is unchanged by mirror reflection. Taking one simple example (see Atkins 1972), a charge moving parallel to an electric current is deflected toward the current by the current-induced electromagnetic field. Under mirror reflection, both the field and charge deflection are reversed. In the mirror, therefore, the charge is still deflected toward the current. What if charge deflection were *not* reversed by the mirror operation? The mirror image charge would be deflected *away from* the current in contradiction of electromagnetic law. Parity would be violated. This does not happen in the case of electromagnetic interactions. However, the Wu experiment, suggested by Lee and Yang, demonstrated that an analogous phenomenon *was* observable for the weak interactions (those involving nuclear decay). Here, the direction of emission of electrons from radioactive nuclei was *not* reversed by mirror reflection as was the electromagnetic field direction. Consequently, while β-emission occurred antiparallel to the field in the unreflected Wu experiment,

in the mirror-reversed counterpart electrons were ejected parallel
to the field. This served as a clearcut violation of parity.

The significance of parity violation for the four-dimensional
space-time continuum constituting our physical universe may be
difficult to appreciate. A simplified two-dimensional analogy may
prove helpful in this respect. Figure 2.1 depicts a two-dimensional
world. It appears as a surface that we must imagine to be built
up from the physical laws of that world. The continuum is
unbroken in that every right-oriented body is matched by its left-
oriented complement or, at least, is matchable in principle, since

Figure 2.1

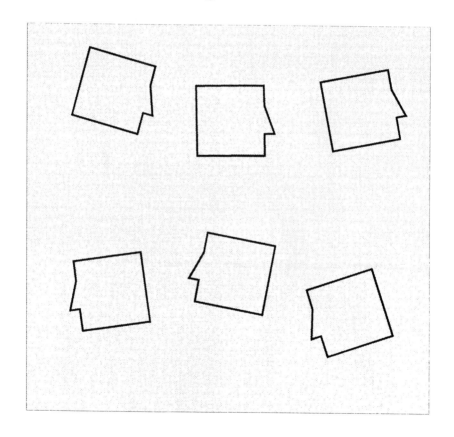

Symmetric world

the mirror images of all bodies are admissible under the laws of the continuum. But suppose we discover an event in this surface whose mirror image contradicts physical law. The image cannot be represented in the continuum. It is pictured in figure 2.2 as a hole in the surface. The two-dimensional space-time continuum is no longer complete in itself. Its symmetry is broken. By analogy, the demonstrated violation of parity in physics leads to the inescapable conclusion that our four-dimensional world is incomplete, asymmetric. The mirror reflection of the Wu experiment constitutes a tear in the fabric from which our universe purportedly is woven.

Figure 2.2

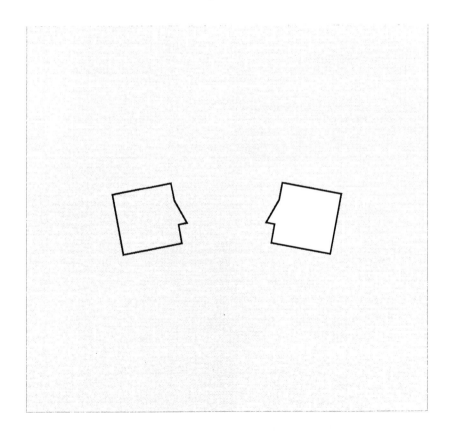

Asymmetric world

Naturally, the primary objective of the physicist is to "mend the tear." Would symmetry not be restored if a more general continuum could be defined that would be capable of accommodating the Wu phenomenon?

As noted at the outset of this paper, restoration of symmetry can be understood in terms of rotation through an added dimension. If reality were strictly limited to the continuum pictured in figure 2.2, hyperdimensional rotation would not be physically possible and asymmetry would be absolute. But let us expand our conception of the physical world by permitting the possibility of a rotation *perpendicular* to the surface. It would then be very simple to transform the right-facing profile embedded in the two-dimensional space of figure 2.2 into a left-facing profile: we would only need to turn the profile over. Here we begin to see a basis for resolving the paradox posed by Wu.

However, if continuity is to be regained in an admissible way, the hyperdimensional operation that would bring this about could not merely be extrinsic to the geometric structure of the surface. This is because the structure in question embodies our physical universe; therefore, any rotation occurring merely *outside* of it could have no real physical significance. Thus, in seeking a continuum to account for the violation of parity, the task becomes one of specifying a geometric structure in which the hyperdimensional rotation that changes right into left is incorporated as an *intrinsic* feature.

THE STRIP OF MOEBIUS AND THE PRINCIPLE OF SYNSYMMETRY

Elsewhere (Rosen 1973 *[and chapter 1]*), I have discussed the scientific and philosophical implications of the Moebius strip. Presently we discover that the strip possesses just those properties required of a structure that would address the problem of parity violation. This can be demonstrated by further developing the two-dimensional analogy. Let the space-time continuum be represented as one side of a two-sided cylindrical band, and imagine that in this continuum, a right-facing profile is embedded *[see fig. 1.2a].*[1] As in figure 2.2, the fundamental asymmetry disclosed by the Wu experiment is symbolized by the fact that nowhere in the continuum does the profile's mirror-reflected counterpart appear. And no movement of the profile in its continuum, no amount of rotation along its side of the two-sided band, will change its orientation. Of course, the orientation of the profile is reversed on the *other side* of the band, but even if we altered our model of reality to include this other side, the structure of the cylinder is such that we would be left without a natural means of making the transition through the higher dimension from one side to the other. The cylindrical ring does not give us a geometrically intrinsic means of restoring the lost symmetry, for it possesses two independent, simply dissociated sides.

Now consider the band of Moebius *[fig. 1.2b]*. In the course of Moebius rotation, the right-oriented profile *is* transported through the higher dimension to the other side of the strip, where it becomes a left-oriented profile; and this process of symmetry restoration (wherein right is superimposed on left) occurs *geodesically*, that is, by following the natural contour of the strip without crossing an edge. The mathematical attribute reflected in the phenomenon is *one-sidedness [as discussed in chapter 1]*. The Moebius model thus suggests the manner in which space-time continuity, broken at level n, is intrinsically regained at $n + 1$.

For all that, the fundamental synthesis of symmetry and asymmetry sought by Lautman still has not fully been articulated. A more general model of space-time is required, one that conserves the dialectical tension of opposites by displaying the breaking of *all* symmetries. We cannot stop with the restitution of symmetry at level $n + 1$. Here too, the development of novel asymmetry must be expected. Parity must fall anew. Each reinstatement of continuity needs to contain the seeds of higher-order discontinuity. Or, to use the language of Weyl, the transformation of space in which the point is mapped into its mirror image cannot be an automorphism; the structure of space must *change*, continuously evolve. Such a dynamization of space-time would fuse symmetry and asymmetry so intimately that the old nomenclature would become inappropriate. We may characterize this integrative conception of space-time as "*syn*symmetric" and acknowledge that the Moebius geometry, in and of itself, does not do it full justice.

However, a crucial property of the Moebius strip thus far has been neglected. It not only restores symmetry by superimposing asymmetric profiles; *it is itself asymmetric*, coming in right- and left-handed forms . . . *[at this point in the original article, I introduced the* Klein *bottle (as in chapter 1): the higher-*

dimensional counterpart of the Moebius strip that fuses oppositely oriented Moebius strips just as the strip had restored the symmetry of the profiles. Then, noting that the Klein bottle is itself asymmetric, *the "hyper"-Klein bottle is intimated—the structure that would restore the symmetry of opposing Klein bottles, and so forth. . .]* Here an infinite progression seems to disclose itself, with the restoration of symmetry ever defining a higher order of asymmetry. The mathematical system that results from the generalization of the sequence: Moebius strip, Klein bottle, hyper-Klein bottle. . .thus provides the basis for a *synsymmetric* notion of space-time.

In concluding, we note that the Moebius-Klein system appears to have an interesting implication for the relation between the point structure of space and its global connectedness. For a preliminary idea of this, let us assume that the profile that we have pictured as embedded in the Moebius strip represents the *point-element* of two-dimensional space, "blown up" to reveal its asymmetry; here the Moebius surface itself, taken as a whole, would model the overall topological connection of that space. What we have found is that the *restoration* of the profile's symmetry, in effect, is associated with an *expansion* of the two-dimensional realm that thrusts the Moebius structure into a new role: it is now the point-element of the *three*-dimensional world whose global connectedness is modeled by the Klein bottle—and that Moebius element is asymmetric, just as the "profile element" of the Moebius-connected world was.

The comments of Gardner are particularly relevant in this regard. He asks:

> is it possible to construct models of the cosmos in which space has an intrinsic handedness?. . .You might think that an overall twist of space, comparable to the twist of a Moebius strip, would do the trick, but it doesn't. The twist has to be present at every point. . . One has to construct a space in which there is some sort of fine, unobservable "grain" which provides a uniform asymmetric twist. . . (Gardner 1964, p. 256)

Although Gardner thus divorces the question of "overall twist" from that of "fine grain," our generalization of the Moebius model enables us to assert that the model of space possessing intrinsic handedness at every point is the very same one exhibiting an overall twist! Hence, the Moebius-Klein geometry describes the manner in which a cosmical twist at level n becomes an infinitesimal twist at $n + 1$.

3

Creative Evolution (1980)

"Who and what am I?" That is the plaintive query so often heard now in our age of alienation. The "consciousness explosion" presently being witnessed in America—Yoga, TM, Zen, encounter groups, transactional analysis, drugs, est, biofeedback, the search for geneological roots, and so forth—attests to the broad and sometimes frantic search for personal identity. And we are no less concerned about our collective *human* identity as we seem to edge closer and closer to the brink of unknown territory. This building sense of urgency about identity lends special significance to the question of human origins and human destiny; the question of *evolution*. A far-reaching reappraisal of this critical issue appears to be needed. Yet, in established circles, the underlying assumptions of the time-honored view have traditionally been shielded from outside scrutiny.

The still generally accepted doctrine is, of course, the one first set forth by Charles Darwin in the nineteenth century. Forms evolve gradually over thousands and millions of years, all changes being wrought in accordance with two basic principles: chance variation and natural selection. That is, nature—the condition of the environment that happens to prevail in a given region—selectively favors particular variations in the local population over others. The favored individuals, owing their superior characteristics to genetic accident, survive and proliferate; those less suited to environmental circumstances eventually become extinct.

But is Darwin's theory of evolution really what it purports to be? By definition, a theory of evolution must account for genuine change, whether rapid or gradual. If it deals with relatively superficial change while upholding changelessness at a more fundamental level, is it not somewhat misleading to call it a "theory of evolution"?

Darwin's conception repudiated the religious doctrine of Divine Creation, which at that time enjoyed wide influence. The world was fashioned by God, said the Church, and in that divine procedure, man was created separately. A true distinction exists between man and all other subjects in the kingdom of God, for man is the only being possessed of an immortal soul. Charles Darwin, on the other hand, affirmed the *continuity* of man with his fellow creatures. For Darwin, the *laws of nature* were supreme and everlasting. He endorsed the idea of "uniformitarianism" formulated by Charles Lyell in the field of geology. This principle suggests that the same natural laws that operated millions of years ago operate unaltered today and will continue to do so indefinitely into the future. Thus, there was no six-day drama of special creation occurring a mere five thousand years ago. Nothing was, or ever is, really created, and therefore nothing can be created separately. Rather, forms "slowly evolve," one from the other, in keeping with unchanging universal laws. Man is simply the most recent manifestation of this continuous process.

Certainly we differ in appearance from the giraffe or jellyfish, and in some ways, behaviorally, we show ourselves to be so much more clever. But the real impact of Darwinism is felt at a deeper level, for beneath the observed differences it presupposes a profound sameness. All transformations of structure and function issue from the invariant principles of chance genetic variation and natural selection. This is the common denominator to which all diversity in the organic family is reduced, the axiom of nature that serves to undermine any notion of basic human advancement. No matter how lofty our aspirations or noble our ideals, Darwinism, in effect, assures us that we have not escaped the "law of the jungle." Then, in the final analysis, can the forces that motivate us be any less base than those that drive our companions on the phylogenetic scale? The Darwinian tenet indeed has been systematically brought to bear on human motivation, and that application has been a dominant factor in twentieth-century thinking. We human beings have been portrayed as biological machines who act blindly and irrationally against our own best interests (as in Freud's conception of man).

Undoubtedly, the influence of Darwinism on our lives has been considerable, and its effects have been explored extensively by a great many thinkers. In this essay, I wish to emphasize only that Darwin's theory of change, when viewed most essentially, is actually a statement of *changelessness.* Surely no one can deny that change occurs, but change can be vitiated and ultimately explained away. Operating within the long prevalent *Weltan-*

schauung of changelessness, the effect of Darwin's theory would be to accomplish just that. Why has this view won such wide approval? Perhaps because it has helped us so masterfully to reduce the cognitive dissonance between, on the one hand, the precepts of classical Greek and post-Renaissance thought we have been obliged to accept, and, on the other hand, our primary awareness that change does take place. By dwelling on the topic of change, by expounding the presumed evolutionary process in elaborate detail, the Darwinist diverts our attention from the fact that he is not actually dealing with *fundamental* change, that change is being trivialized. Thus our conflict is apparently resolved. We manage to convince ourselves that we are understanding change, yet we do not find it necessary to question the paradigm of changelessness that has shaped and restricted our thinking for so many centuries.

THE MOEBIUS-KLEIN SYSTEM:
A SYNTHESIS OF CONTINUITY AND DISCONTINUITY

But what if the conceptual "sleight of hand" does not succeed? Suppose we just cannot convince ourselves. Then there is no alternative but to raise the appropriate questions, however unthinkable they may be within the established framework. We are seeking an image of change, of *creative* evolution. Uniformitarian monism, therefore, will need to be challenged. Nor will the old religious doctrine of Divine Creation suffice. No being is ever subject to remote control. Contemporary critiques of Platonic and Cartesian dualisms make it clear that an entity will not be influenced in the slightest by an agent acting entirely from without.

What seems required here is a *monistic dualism*, to echo the theme of Nahum Stiskin (1972). This suggests a fusion of continuity and discontinuity. Evolution must entail the "radical discontinuity" spoken of by Pearce (1973). "This discontinuity in the growth of mind makes ridiculous our current attempts to equate man with the lower animals" (1973, p. 144). But all transformations must be of an *internal* order. Contrary to the principle of uniformitarianism, the laws of nature themselves must change, only the operation must be of an inner, *self*-transforming character. On the whole, we could say that while man is not detached from nature, he embodies natural laws that are qualitatively distinct from those of his fellow creatures.

The extant set of natural laws compose the fabric of a manifest universe, a given domain of space and time. The radical discontinuity through which the laws of nature mutate should therefore involve a transfiguration of a space-time framework. The continuity of the particular space-time regime should be broken and a new regime ushered in. Yet the generative process must somehow be achieved *continuously*, without the Cartesian leap into void.

In several other contexts *[see chapters 1, 2, and 8]*, this seemingly unfulfillable requirement in fact has been met. My solution calls for an active use of geometric imagination.

[Once again, the Moebius principle is articulated, this time in the context of offering a general concept of phylogeny.]

CREATIVE EVOLUTION AS A HIERARCHY OF REFLECTIONS

Soon I shall attempt to explore creative evolution more specifically. Four evolutional epochs are to be considered. The guiding image throughout will be the mathematical system described above. The epochs shall broadly be conceived as members of the extended Moebius-Klein family. But if we are fully to appreciate the workings of metamorphosis, we must first examine one more general characteristic of creative evolution: The Moebius-Klein system constitutes a *hierarchy of reflections*. That is, the operation by which novel space-time domains are geodesically produced is a process of mirroring.

The reflexive quality of Moebius movement can best be grasped by studying the distinction between a part of a whole and an *aspect*. *[The distinction in question is brought out in chapter 1 by means of the Necker cube (fig. 1.4a). Since each perspective of the cube completely encompasses the other perspective and the cube in its entirety, individual perspectives are not merely isolated parts; instead they are regarded as* aspects. *It is clear that opposing sides of the one-sided Moebius surface are connected to each other in the very same way.]*

This intimate interconnectedness of sides makes it evident that the aspect does not incorporate something that is foreign to it. Clearly, opposite sides of the Moebius strip, which express the opposition of left and right, are never simply separated but are bound up internally as one. We therefore can say that left and right—and generally, subject and object, man and universe—

rather than being viewed as merely dichotomous, should be understood as "alter-egos" of one another. The Moebius embrace of the world is an embrace of self as well. The objective, apparently external world in this way becomes subjectified—I come to know it as my self. Concomitantly, the old subjectivity of my ostensibly isolated self becomes *ob*jectified, in that I can turn back upon and observe this more limited mode of my being from the vantage point of the whole.

Nevertheless, the efflorescent character of the Moebius-Klein system never permits one to rest on one's laurels, nor even to garner them in full. The expansion toward self-awareness that promises to close the Moebius circuit of identity leaves off at the very threshold of fulfillment. Once there, because the higher "Klein-bottle identity" has been incubating all along, we find, not "man supremely wise," but "superman," supremely *ignorant* and ready to embark on a new, elevated path to self-enlightenment. The occurrence of each reflexive self-embrace implies the passing away of the being in the process of reflecting, and contains within it the developing basis for the next order of reflection. So the hierarchy unfolds.

There is a remarkable illustration of the type of expansion found in the Moebius-Klein system. I am referring to M. C. Escher's graphic work *Print Gallery* (fig. 3.1). A boy is shown standing in an art gallery gazing at a print. In viewing Escher's picture, which we will call the "master print" (to distinguish it from the prints depicted in the gallery), we observe the boy. He appears to be embedded in the two-dimensional world of the master print, while we experience the three-dimensional field lying outside. However, Escher has provided him with a means of departing from his two-dimensional frame of reference and aligning his consciousness with our own.

Let us first review Escher's own commentary on *Print Gallery*:

> We have here an expansion which curves around the empty centre in a clockwise direction. We come in through a door on the lower right to an exhibition gallery where there are prints on stands and walls. First of all we pass a visitor with his hands behind his back and then, in the lower left-hand corner, a young man who is already four times as big. Even his head has already expanded in relation to his hand. He is looking at the last print in a series on the wall and glancing at its details...Then his

Figure 3.1

Print Gallery
© 1956 M.C. Escher/Cordon Art-Baarn-Holland.
Collection Vorpal Gallery New York (Soho)/San Francisco.

eye moves further on from left to right, to the ever expanding blocks of houses . . . and this brings us back to where we started our circuit. The boy sees all these things as two-dimensional details of the print that he is studying. If his eye explores the surface further then he sees himself as a part of the print. (Escher 1971, p. 16)

Escher's expansion suggests Moebius rotation by exhibiting the manner in which a detail of the whole encompasses the whole.

If we take the liberty of supposing that the row of prints depicted by Escher is a single, movable print, we see how the one represented two-dimensionally at the right end of the gallery rotates clockwise, expanding and taking on three-dimensional perspective. In this way, the print manages to "swallow" the master print, symbolizing the microcosm's embrace of macrocosm. When the boy in the picture follows this movement, he will come to view himself as we three-dimensional beings view him, as a circumscribed detail of a two-dimensional canvas.

Is this not the meaning of topological "one-sidedness"? In the case of the Moebius strip, a side twists through the higher dimension to envelop the other side and so incorporate the whole; the two sides are one side, the part, as an aspect, is fused with the whole.

But let me propose an amendment to Escher's interpretation of *Print Gallery*. By following the clockwise expansion, the boy's consciousness is dilated so that he is able to reflect upon himself, to view himself as an object—"he sees himself as a part of the print." Yet we may ask whether, in the end, it is the boy *as such* who contemplates himself. When the expansion is completed, is the boy in the picture being examined through the subjective eyes of the boy in the picture? Evidently not. We may first observe that his previously limited subjectivity has been universalized, widened by the awareness he has gained of his two-dimensional world in its entirety. But metamorphosis has taken place beyond even that. With the expansion of consciousness consummated, a higher subjectivity has been called into being. The observer of the master print, no longer an occupant of that two-dimensional realm but now situated in the *three*-dimensional domain, is not simply the boy we initially encountered. In coming to view himself as object, to *reproduce* himself by self-reflection, he is transformed. The final result is that the boy we observe becomes identified with *us*, the observers. In short, the twist into the higher dimension suggestive of the Moebius principle is as much a turning away from the lower domain as a reflexive turning back upon it.

Moreover, the existence of a higher-order "master print" is strongly connoted. For when the boy's consciousness is stretched over ours, when his identity becomes our identity, do we not find ourselves confronting circumstances similar to those he faced? Is there not an implicit progression here compelling us to imagine our own space and time being blended on a "hyper-canvas" that we will come to know by performing the appropriate expansive reflection? Underlying Escher's provocative creation is thus the open hierarchy of space-time domains of the Moebius-Klein system. At each level, a reflection occurs that accompanies the action of encompassing a macrocosm, of attaining aspecthood. Reflections accumulate continuously, every novel identity predicated upon those that came before, even though each is also *discontinuous* from its antecedents, being the manifestation of a new regime of space and time. Our Moebius-Klein generality is, therefore, a hierarchy of reflections or mirrorings in which identity becomes progressively more convoluted through higher-order reproductions of self. This spiraling synthesis of continuity and discontinuity is the essence of creative evolution.

CONCEPTS OF CREATIVE EVOLUTION

Consider theoretical biologist Howard Pattee's (1972) concept of "self-simplification." He compares the Darwinian view of evolution with his own less conventional notion:

> I would like to propose that self-simplification is the key process in all so-called novelties or archetypes in evolution—also called transpecific or macroscopic evolution—as contrasted to local optimization for which mutation and natural selection appear to be a reasonable mechanism. Self-simplification does not compete with natural selection, but is complementary to it. That is, natural selection works well on relatively simple systems with the result that they accumulate genetic information and hence phenotypic complexity. This complexity is multiplied by the population's growth and stronger coupling with the ecosystem. Consequently, the organism reaches a level of complexity where natural selection stagnates and can only produce nonselective genetic drift. At this stage the more global process of self-simplification can introduce a new hierarchical level of description which selectively ignores the trapped details at the lower level. Natural selection can now begin to operate effectively in the language of this new simplified level; and the process can now repeat with no obvious limit to the level's structure. (Pattee 1972, p. 39)

Simplification is much like classification: "the grouping under a single class of objects which, according to a more detailed classification, fell into more than one class, i.e., the objects were initially further distinguishable" (Pattee 1972, p. 37). Therefore:

"all symbols and records, insofar as they are classifications of events, are consequently simplifications of these events since they selectively ignore some details" (1972, p. 37).

The *genetic* record is of primary interest to Pattee. The evolution of life from the inorganic depends on the emergence of a classification or description of events that is distinct from the events themselves. In other words, life depends on the separation of a genotype from phenotypic being. Living systems are defined as systems that can reproduce themselves, and self-reproducing units must possess self-descriptions from which the reconstruction can be carried out. "Then it is fair to say that life originated by self-simplification of natural events" (Pattee 1972, p. 37).

Put another way, the first step in "transpecific" evolution occurred through *reflection*. Prior to the advent of life, no true individuality existed, for the "individual being" was a mere detail, a localized object in a phenotypic world. There was no descriptive simplification of macrocosmic events apart from that macrocosm. Then a reflection occurred in which "the underlying physical and chemical events" of the macrocosm could be described by an individual being, that is, *mirrored in the microcosm*. With the outer world thus recapitulated internally, self-reproduction became possible.

Clearly, the separation of genotype from phenotype implies transcendence of the given phenotypic regime. If classification is to occur, the classifier cannot be subject to the same constraints as the events he classifies. To describe a system, one must remove oneself from it to gain perspective; description necessarily involves freedom of movement in a higher dimension. Though Pattee did not adopt a specifically geometric frame of reference in his illuminating discussion, a closer examination of the appearance of the life function provides a clearcut demonstration of the implication of higher dimensionality.

Clifford Grobstein (1973) recently introduced the idea of "neogenesis," the emergence of novel properties at higher levels of organization that arise from, but are not entirely determined by, lower-order circumstances. Grobstein's case in point is the process of *folding* in the protein molecule, a process Pattee characterized as "perhaps the most fundamental example of primitive hierarchical organization" (Pattee 1973, p. 90). The emergent property is that of the enzyme, a function vital to life. "Without enzymes," says Pattee, "there would be no DNA replication, no transcription to RNA, no coding, and no synthesis" (1973, p. 99)—that is, no life. According to Grobstein, the enzyme function appears, in effect, as a hyperdimensional identification of local areas that are widely separated in one-dimensional space.

In Grobstein's illustration, we see how a linear chain of amino acids (these are the basic constituents of the protein) is folded into a three-dimensional configuration called a "native protein." The operation (depicted in fig. 3.2) depends on the strategic location of certain chemical groups associated with the amino acids:

> Some of the amino acids along the chain have side groups which are hydrophilic and some hydrophobic. The general tendency in water solution is for the hydrophobic side groups, the ones not readily miscible with water, to move into the interior. The hydrophilic ones, on the other hand, tend to remain at the surface. The difference between the hydrophobic and hydrophilic distribution along the chain is an initial generating force to the folding of the molecule. (Grobstein 1973, pp. 38–39)

The significant fact about the folded structure is that:

> In this "native configuration" a critical "new" property appears— an enzymatically active site [the cross-hatched region of fig. 3.2] has been formed which was not present previously. Note that the site consists of several local areas of the linear order which are widely separated in . . . the unfolded . . . configuration, but are brought together in the so-called native one . . . The linear distribution of amino acids does not by itself produce enzymatic activity—suitable folding is necessary for the activity to *arise*. (Grobstein 1973, pp. 40–41)

Grobstein's observations support the idea that qualitative evolution is an emergence of higher dimensionality. The work of philosopher Oliver Reiser (1966) lends additional credence to this theme. It is likely that Reiser would have agreed with Pattee about self-simplification. In fact, the former was led to the early conclusion that "there are always two kinds of simplicities in nature: *simple simplicities* and *complex simplicities* . . . [E]very entity is a simple simplicity when treated as a component of a higher aggregate and a complex simplicity when regarded as an emergent from its antecedents in the . . . evolution of matter" (Reiser 1966, p. 346). The enzyme function that emerges through "neogenesis" is evidently a "complex simplicity" in relation to

Figure 3.2. *Neogenesis.*

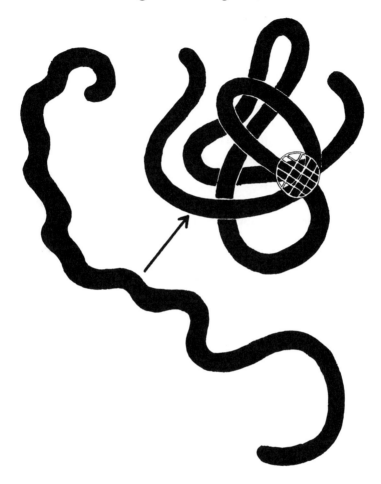

Schematic drawing showing the conversion of a linear chain of amino acids into a three-dimensionally configured native protein. During the process of folding, amino acids which are widely separated in a linear sense are brought into spatial proximity to form an active center (cross-hatching) for the emergent enzyme function.
(From Clifford Grobstein in Howard H. Pattee's *Hierarchy Theory* © New York: George Braziller, 1973.)

the linear form of protein; it is the entity so crucial to the effective separation of genotype from phenotype and, thus, to the production of life. But note that Reiser's broad conception implies that the enzyme is, at the same time, a simple member of the higher-order system that becomes stabilized once life is established, that is, the enzyme is also a "simple simplicity."

Unlike Pattee, Reiser strongly emphasized that the hierarchy of simplifications is one of emergent dimensionality. Throughout his career, he recognized the necessity of enriching the mathematical image of the cosmos, revising it from a static, one-level Euclideanism to a geometry that would permit genuine metamorphosis. In his last major work, Reiser intimated that a hierarchy of *Moebius surfaces* might be required to account for the transition from one dimension to another. As he put it, nature might " 'round the curve of dimensionalities'. . . through ascent into the higher dimensions by way of a generalized concept of Moebius' surfaces" (Reiser 1975, p. 183).

THE FOUR KNOWN LEVELS OF CREATIVE EVOLUTION

According to Reiser and to convention, the three domains manifested within the range of human observation are the physical, the biological, and the psychological. Actually, *four* levels may be identified. The biological function has been found to be a hyperdimensional reflection upon the physical. A description separates from that which is described, producing an *individualized* form of being (i.e., a being in microcosm) by the folding of universal being (i.e., the being of macrocosm). However, the simplest living entities have no individuality *in themselves*. At this most primitive level, each and every genotype is a description of the same phenotypic truth; the microcosmic reflection upon macrocosm is identical from one microcosm to another. With individuality only serving to mirror the universal, there is no individuality per se.

In this regard, the writings of Jules Lachelier (1962) are of interest. It seems that this nineteenth-century French philosopher also viewed evolution as a hierarchy of reflections. "Will is the principle and the hidden foundation of everything which exists," said Lachelier (1962, p. 155). At the core of reality is the "will to be." In life, the universal will to be is reflected as an individualized "will to live." If this first reflection characteristic of simple life is called a "doubling" of the will to be, we can agree with Lachelier that higher forms of life "redouble it, as it were, and reveal it to themselves in their affective states" (1962, p. 155). Thus a second order of reflection is described. The higher forms of subhuman organisms are distinguished from the lower by their capacity for complex sensory experience. Particular feelings crystalize around particular perceptions and these feelings or affective states express the individual's will to live. At this stage, individuality does exist for itself. Each microcosm is not merely an "assembly-line" mirror image of the macrocosm; it reflects the distinctive aspects of that macrocosm defined by its own experiences. We have then not just the will to live but the will of a *particular being* to live, a being through whose feeling-toned sensations the macrocosm takes on a more differentiated form.

Human being is a further advance in the global evolutionary process; individualization reaches a higher level in man. Lachelier commented that some subhuman creatures can partially detach themselves from their sense impressions and are able to "see them float before them like a sort of dream" (1962, p. 155). Man separates himself much more completely and so can reflect upon his impressions of reality. This third order of reflection endows the will to be with objective status. The world is brought into focus as a space extended before the subjectivity of man. Its existence is established as real and true by the being capable of reflecting upon it. The world—that is, the will to be—is hence transformed once more.

We may summarize the four levels of "transpecific" evolution:

1. macrocosmic being: the *physical* level
2. the macrocosm reflected in microcosmic images of itself: the *biological* level
3. the individualization of the microcosm through a second order of reflection: the *affective* level
4. the further individualization of the microcosm by a third order of reflection that objectifies the macrocosm: the *psychological* level

In concluding, it may be interesting to note that the four phases of creative evolution identified here appear to be paralleled in two *esoteric* traditions. In the "occult science" of Rudolf Steiner (1861–1925), an attempt is made to extend the scope of scientific activity beyond the external world to the inner landscape of the mind. It was Steiner's conviction that consciousness is the hidden source of reality and that certain disciplined meditational practices can lead to a transcendent, "supersensible perception" of that reality. "For supersensible perception," said Steiner, "there is no such thing as 'unconscious,' but only varying degrees of consciousness. Everything in the world possesses consciousness" (1972, p. 135). In his study of world evolution, Steiner described four "degrees" of consciousness that are akin to the hierarchy of reflections outlined above. The "dullest" form of awareness is the *physical*. Next in order is the *etheric*, associated with the life function. The third degree of consciousness is the *astral*, feelings and sensations experienced in an immediate way by the higher forms of animal life. Steiner spoke finally of *egoic* consciousness, which the human being possesses exclusively. It is through the ego that the human not only senses his environment here and now but can reflect upon these experiences, relating them to other places and other times.

The same four phases of evolution seem derivable from a tradition much older than Steiner's. The Chinese sages of old, says Chuang Tsu:

> took as their starting point a state when the existence of things had not yet begun. . .The next assumption was that though things existed, they had not yet begun to be separated. The next, that though things were separated in a sense, affirmation and negation had not yet begun. When affirmation and negation came into being, Tao faded. After Tao faded, then came one-sided attachments. (Jung 1955, p. 100)

At the physical level, "the existence of things had not yet begun" because there was no individuality, only the undifferentiated macrocosm. The microcosm came into being in the rudiments of life, but simply as a mirroring of the macrocosm: "though things existed, they had not yet begun to be separated." With the development of complex sensation and affective awareness, microcosmic being no longer was a mere reflection of the macrocosm but existed for itself. Thus "things were separated" although "affirmation and negation had not yet begun." The fourth assertion of the sages was a comment on the human condition. We affirm our existence by objectifying our particular world; other possibilities are negated. The sages of old were observing that when this is done, we forget the primordial ground from which our world derives as but one of an infinity of possibilities.

It is now appropriate to reiterate the major theme of creative evolution: the unity of continuity and discontinuity or, we may say in these closing passages, of *being* and *becoming*.[1] In the broadest terms, this signifies the preservation of being across levels of becoming in the Moebius-Klein hierarchy, but the synthesis also suggests that being *at any given level* cannot rightly be separated from the process that yields the subsequent level. Being at level n is not *followed* by the becoming of $n + 1$; rather, the becoming is an integral aspect of the being. In the present context we may say that physical being is intimately interwoven with biological becoming, biological being with affective becoming, affective being with psychological becoming, and psychological being with an order of becoming still unclear to us, since the becoming now at hand is *our own* and is not yet complete. In each case, the being who reflects is simultaneously undergoing a higher-order reflection. Thus, the $n + 1$ mode of being incubates amid the ongoing existence of n.

4

The Concept of the Infinite
and the Crisis in Modern Physics (1983)

This essay is predicated on the assumption that the conceptual difficulties besetting modern theoretical physics are serious enough to warrant a thoroughgoing reexamination of its foundations. I intend to show that the idea of *infinity* is central to these difficulties. My effort is motivated by the conviction that we must clarify the role of the infinite in scientific analysis if we are to come to grips with the challenge that faces us.

INFINITY: OPEN AND CLOSED

Conventional wisdom tells us that an infinite number or amount is one that is indefinitely great. Another term is always available beyond the last term counted in such a sequence. This kind of infinity therefore may be called an *open* or *potential* infinity. Almost three hundred years ago, with the advent of the calculus, the method for *closing* or *actualizing* the infinite appeared to have been produced. But in the last century, the development of set theory led to a clarification of the concepts of convergence to a limit and actual infinity.

According to set theory, if we are to understand how infinity can be actualized, we must revise our thinking about groupings of elements. An element is considered a member of a group if it possesses certain characteristics and not others. By shifting our attention to the characteristics themselves and away from the elements characterized, we form an understanding of the group that transcends any particular experiences we might have had with its individual members. As Kramer illustrated, "We can speak of 'Americans' without having a personal acquaintance with each citizen of the United States" (Kramer 1970, p. 579). This transition from the concrete, open-ended operation of counting elements one by one to a conception of them in the more abstract context of membership in a class or set enables us to appreciate actual infinity. Though we cannot count the members of an infinite set, we can still confirm that the set is infinite by verifying that it possesses a certain characteristic, namely, that it can be made equivalent to a proper subset of itself. (This quality is demonstrated if the well known principle of one-to-one correspondence between set and subset is satisfied.)

Then actualization of the infinite clearly entails a *qualitative* transformation. In the words of Georg Cantor (nineteenth-century set theorist and innovator of the idea of the transfinite), the goal of the infinite is obtained by "stepping out of the series" (Cantor 1976, p. 88). Any attempt to build up an infinity from below by mere iteration will be an exercise in futility.

How did Cantor specifically deal with the problem of achieving the infinite? The sequence of the positive real whole numbers, 1, 2, 3, . . . , v, . . . , is generated by

> repeated positing and uniting of basic units. . . , the addition of a unit to an existing, already formed number. I call this. . . the *first principle of generation*. The number. . . of the numbers v of class (I) to be formed in this way is infinite, and there is no greatest one among them. As contradictory as it would be, therefore, to speak of a greatest number of class (I), there is, on the other hand, nothing objectionable in conceiving of a *new* number—we shall call it ω—which is intended to be the expression for the fact that the totality (I) as a whole be given in its natural and lawful succession (similar to the way in which v is an expression for the fact that a certain finite number of units is unified into a whole). . . if any definite succession of defined whole real numbers is given of which there is no greatest, then on the basis of this second principle of generation a new number is created. (Cantor 1976, p. 87)

In examining Cantor's discussion of the second principle of generation, one is unable to find a basis for actualizing the infinite (in the illustration, the appearance of the transfinite number, ω, beyond the finite sequence, v) other than the mere existence of the *un*actualized, that is, potential infinity. The effect is a tacit assertion that if there is an open infinity, this of itself is the necessary and sufficient condition for its closure upon the transfinite! Does this not amount to *invoking* the actual infinite, defining it into being by fiat?

There can be no denying that to attain the infinite, we must "step out of the series"; there will be discontinuity in the given number field. But discontinuity alone, an absolute discontinuity, signifies nothing more than an incompleteness in understanding.

The challenge is to discover the *deeper* continuity operating to engender the infinite. The continuous transformative process by which the infinite is realized must be brought to light.

For the time being, I would like to point out that the problem of generating infinity is not genuinely addressed in conventional mathematical analysis, that is, in the standard calculus with its method of limits. But we should be able to see better how the issue is circumvented there, by first considering the relative nature of the infinite.

THE FIRST PRINCIPLE OF THE RELATIVITY OF THE INFINITE

The relativity of the infinite is implied in Cantor's conception of the transfinite number, for Cantor emphasized time and again that such numbers are not vague but *definite*, a term consanguineous with *finite*. It is precisely because we ordinarily take the infinite to be absolute that our minds boggle at the notion of a number that is "infinite" yet "definite." The difficulty is resolved as soon as we qualify our use of language. The transfinite number, ω, is infinite *relative to* the lower-order sequence, 1, 2, 3,..., ν,.... Yet in relation to the sequence of which *it* is a member, $\omega + 1$, $\omega + 2$,..., $\omega + r$,..., ω should be regarded as *finite*. Mathematician Charles Muses (1975a) makes the same case from a geometric perspective:

> Now it is known that in a given n-dimensional space, the operative presence of an extra or higher dimension that is not suspected, can reflect itself effectively in the mathematics of the lower dimensions as an infinity. Thus a finite unit square (second dimension or D_2) may be considered as an infinitely long line... finite area (D_2) reflects as an infinity when only a linear continuum (D_1) is considered. Similarly, finite volume (D_3), when restricted to only two dimensions, may appear as infinite area (D_2). Thus, in general, we have the mapping D_{n+1} ∞ D_n. where the left-hand member is some finite portion of (n + 1)-dimensional space. (Muses 1975a, p. 27)

Given the domains of analysis n and $n + 1$, given that $n + 1$ is viewed as an open infinity from within the constraints of n, and given that the transition from n to $n + 1$ closes the infinity, finitizes that which was infinite: How does the classical method of limits operate? What does it accomplish?

The method of limits does *not* provide the transition from *n* to *n* + 1, does *not* close the open infinity, does *not* finitize that which was infinite. For the method proceeds not from below but from *above*, as it were. In this regard, Cantor spoke of the relationship between the lower-order series, α_ν, and the transfinite number, *b*:

> It is not at all true that the number *b* is defined as the "limit" of the numbers α_ν of a fundamental series (α_ν). This would be a logical mistake . . . [O]ne need not first *obtain* it [i.e., the number *b*] by a limiting process but on the contrary—through its *possession* one is convinced of the feasibility and evident admissibility of the limiting processes. (Cantor 1976, pp. 82–83)

So the standard method begins with the infinite already in hand, that is, finitized. Geometrically speaking, we may say that the process commences from the *pre*actualized, higher-dimensional regime and works backwards from there to the lower domain. By following this procedure, an interdimensional analysis is performed that yields a quantitative description of motion or change.

The simplest application is illustrated by the ancient philosophical problem of the dichotomy posed by Zeno. A runner is to traverse an 80-meter race course. Before reaching his objective, he must travel 40 meters. Similarly, to gain 40 meters he must move 20, to advance 20 he must pass through the 10-meter point on the continuum that separates him from his goal, and so on. If this progression is continued, we find the runner unable to take even a single step towards the finish line, much less approach and cross it.

Zeno's paradox of the dichotomy suggests the impossibility of motion. It was founded on the assumption that a line is infinitely divisible, an infinity of points lying between the runner and any other location. However close is point *P* to *P'*, this mathematical presupposition appears to create the effect of a void. The proximity of *P* and *P'* may be arbitrarily increased, but points will not converge.

We recognize the difficulty at once. The Zenoan infinity is a *potential* infinity, and we have seen that this potential can never be actualized while continuing to operate numerically within the lower order of finitude. Naturally, in practice, the runner *can* traverse the course, just as we can draw a line on a sheet of paper by extending a point. The standard analysis of this point-to-point motion begins when the motion is a *fait accompli*, that is, begins from the *higher* order of finitude, namely, the line segment that connects the points. However short this finite interval (and, proportionately, the time it took to draw it), it will be the completed expression of the total change in position that has taken place. But by the mathematical procedure of *differentiation*, we can obtain the *derivative* of position, that is, the *instantaneous* rate of change in position from which the completed motion derives. Differentiation is achieved by the *method of limits*. For the finite condition, change in position is termed $\triangle Y$ and is taken as a function of time elapsed, $\triangle X$. By taking $\triangle X$ to be arbitrarily small so that it approaches zero (the infinitesimal) as a limit, the instantaneous expression of change in position with respect to time is:

$$dY/dX = \lim_{\triangle X \to 0} \triangle Y/\triangle X$$

Though $\triangle X$ converges upon zero, the *ratio dY/dX* will have a finite value. Therefore, conceptually, the derivative may be regarded as a finite expression of the higher dimension (D_1) operating within the lower dimension (D_0). It is in this sense that the method of limits "finitizes the infinite," for we have seen that from the strict standpoint of the lower dimension, the higher dimension would appear as an infinity.

Thus we may say that motion transpires in the domain *between* dimensions—between line and point in the case of linear functions—and the method of limits, converging backward from the higher order of finitude, provides determinate access to this domain. The derivative, by revealing the working of the line within the point-position, gives a precise account of movement from point to point. (Of course, higher-order derivatives may be obtained to account rigorously for more complex forms of movement.)

Historically, the standard calculus was highly successful, being an indispensable tool for the analysis of motion upon which the Newtonian universe was built. The method purportedly deals with the problem of infinity, but, to summarize what I have attempted to indicate in this section, its viability is wholly dependent on the prior, thoroughgoing relativization of the infinite. One does not *obtain* the infinite by the method of limits, to cite Cantor's thinking again. The method can succeed only when one is operating in a domain where the infinite is already at one's disposal, has already been finitized. It is within such a domain that the *first principle* of the relativity of the infinite applies. Here we are dealing with a weakened or *absolutely* relative infinity. The conventional mathematical procedure does not come to grips with the question of *generating* infinities, the problem of *building* bridges between dimensions. It merely uses the bridges already erected.

MODERN PHYSICS AND THE FAILURE OF THE FIRST PRINCIPLE OF THE RELATIVITY OF THE INFINITE

Now let us suppose that a bridge has *not* been erected. Suppose that in attempting to describe a dynamic pattern with respect to a given space, n, there is no recourse to a completed, already finitized $n + 1$ regime. Then it would not be possible to show the instantaneous working of said pattern in space n. A finite derivative of the dynamism could not be obtained; the motion with respect to space n would be irremediably divergent, manifesting itself as an infinity. Naturally, determinate analysis would be ruled out. Analytical continuity would yield to fundamental discreteness, since the dynamism would be undifferentiable, indescribable as a continuous function. In such a case, Zeno's kind of paradox, presumably solved once and for all by the method of limits, would find new life, for here infinity would be powerfully revitalized.

This is precisely the dilemma that has arisen in the twentieth century, with science's attempt to penetrate the realm of subatomic process. In the microcosm, there is primary indeterminateness, inherent discreteness—the problem of infinity returns with a vengeance. Mathematician Muses asserted that the troublesome infinities that turn up in quantum theoretic computations are "the predictable result of omitting a higher dimension," and he cautioned that " 'higher dimension' here means not merely a space dimension but a *new* dimension" (1975a, p. 27). The epistemological significance of Muses's distinction cannot be overestimated. In the language of the present exposition, a mere "space dimension" is one to which a bridge already exists, one that is at the disposal of the analyst, an infinity that has been closed or finitized with respect to the analyst's perceptual/cognitive repertoire. This is what makes determinate analysis possible.

By contrast, the profound epistemological challenge posed by the "dimension" of quantum phenomena derives from its "newness," that is, its *embryonic* status. I propose we regard it as a space that has not been fully incubated, an as-yet unbridged infinity, the bridging of which would entail a *transformation* of the analytical repertoire. This interpretation does not seem an unreasonable alternative to viewing the microcosm as a fully fledged space, which clearly it is not, or merely declaring it a "nonspace," which is tantamount to proclaiming its noninterpretability. Moreover, the suggested interpretation is supported in at least a preliminary way by the meta-theory of dimensional generation adumbrated by Muses (1975b), and independently—through the medium of intuitive geometry—by myself (Rosen 1976). (The theory is currently under development *[see chapter 5]* and will be briefly considered in the concluding section of this paper.)

Nevertheless, most working physicists feel it is unnecessary to interpret the quantum domain in this fashion, indeed, to interpret it at all! If the issue is pressed, they can scarcely deny the extraordinary character of that regime, but the prevailing attitude is that speculation on the nature of quantum reality is best left to the philosophers. Of course, for the program of quantum physics to be carried out, a description of the subject matter is necessary, a formalism must be chosen. The impression is created that the mathematical systems developed to account for quantum processes are pure abstractions devoid of interpretive content. Yet this is contrary to the fact. Far from being interpretively neutral, the algebraic structures of extant quantum field theory rest upon a definite assumptive base and the ground selected is the *classical* one. The underlying presupposition is that quantum space possesses the same basic character as the already finitized infinity, a completed, simple continuousness through which we may follow the standard analytic operating procedure, employ the conventional calculus. Is there not a contradiction here? How can essentially nonclassical phenomena be treated in classical terms? By "proper extrapolation to meet the occasion," we are given to believe.

Consider a central feature of the quantum theoretic formalism: analysis by probability. The probabilistic program does not furnish a positive alternative to classical determinism; on the contrary, it institutionalizes its failure. The failure of determinism means the failing of simple mathematical continuity, as philosopher Milic Capek demonstrated in advancing his argument that the constructs of quantum physics raise serious "doubts about spatiotemporal continuity" (1961, p. 223).[1] To Capek, it was obvious that "the concepts of spatial and temporal continuity are hardly adequate tools for dealing with the microphysical reality" (1961, p. 238). But it is just this continuity that the probabilistic approach seeks to retain, and the price paid is exorbitant. Take, for example, the inability to fix the position of a particle in microspace. Habits of macroworld thinking make it difficult to entertain the explanatory proposition that the microworld particle is just not simply located (that is, it does not occupy a single position in space at a given moment), which would imply that microspace is not simply continuous (nonlocality is incompatible with differentiability, a defining property of the simply continuous space). Instead, a *multiplicity* of simply continuous "spaces" is axiomatically invoked to account for the "probable" positions of the particle—"it" is "here" with a certain probability or "there" with another.

Now Charles Muses has cautioned about playing "superficial games with axioms, assuming any self-consistent set that we please." Citing Kurt Gödel's defeat of the formalist's program (spearheaded by David Hilbert) to establish total self-consistency in systems that are devised arbitrarily, Muses says: "We...cannot simply invent axiom games at will." It is Muses's conviction that such exercises in what he calls "sterile abstractionism" do not "go unpunished by contradiction" (1977, p. 72). We may ask accordingly, is simple continuity truly preserved by the method of formalistic invocation?

Each subspace of the multispace expression (the "Hilbert space") is made simply continuous within itself to uphold the mutual exclusiveness of the alternative positions of the particle. Such subspaces must be *disjoint* with respect to *each other*, their unity being imposed externally, by fiat, rather than being of an internal, intuitively compelling order. Thus, in the name of maintaining mathematical continuity, a rather extravagantly *dis*continuous state of affairs is actually permitted in the standard formalism for quantum physics, an indefinitely large aggregate of essentially discrete, disunited spaces.

Of course, this is just another way of indicating the impotence of classical analysis in the face of revitalized infinity.

The quantum level nonlocality in space n may be viewed as infinite motion in n, and this, in turn, can be regarded as symptomatic of the higher, $n + 1$ motion upon n for which a finite derivative is unobtainable by our customary procedure. That procedure would set simply continuous space against discrete existence so as to analytically reduce (differentiate) it. But, to reiterate the proposed interpretation of the problem, the requisite order of simple continuity is not at the analyst's immediate disposal, for $D_n + 1$ is a "*new* dimension," not a mere space dimension. Confronted with novelty and intent on avoiding it, one may invent artificial means to mask its appearance. Thus, we have the "Hilbert-space" invention, the officially sanctioned stratagem for sidestepping a classically irreducible spatial infinity that implies novel dimensionality.

The physicist Steven Bardwell commented that "the failure to deal with . . . built-in discreteness is the source of the manifold contradictions in quantum field theory" (1977, p. 6). Having examined one aspect of the quantum theoretical contradiction (arising when space would be set against discrete existence, that is, when discreteness would be spatially absorbed by supposing a simply continuous quantum space), we turn to the second. Here focus is on the classically continuous functions of D_n. The hope would be to *bypass* the irremedial discreteness indicative of $D_n + 1$, but we find to our dismay that we cannot—discrete existence effectively and implacably sets itself against space. Bardwell identifies this as the case "where the relation between the discrete properties that a source introduces into an otherwise continuous field quantity can only be postulated" (1977, p. 6). For instance, when one tries to solve the quantum field equations so as to determine the interaction of an electron with its own electromagnetic field, this "self-energy" value of the electron turns out to be *infinite*. Physicists generally acknowledge that the appearance of such infinities gives clear evidence of the inadequacy of the standard formalism. Of course, an infinity must be "finitized" if theoretical parameters are to be recoverable by laboratory observations. In the absence of a mathematically natural program for carrying this out, the task is performed by "brute force," by the purely ad hoc calculational procedure known as "renormalization."

Thus, from this second perspective, we again see the primacy of discreteness, of infinity, in the microworld. In the microcosm, continuity cannot exist indifferently in the face of discrete elements imposed from without, any more than discreteness can be legitimately suppressed by a continuity that is formalistically invoked. Perhaps the problem is more obvious when discreteness is allowed to express itself in the particulate mode, the vantage point from which we witness the destruction of otherwise continuous space by particulate infinity. Nevertheless, from the other perspective, where the attempt is made to absorb discreteness into spatiality, we find discreteness arising from within—the spatial infinity develops.

THE GENERAL THEORY OF RELATIVITY
AND THE PROSPECT OF A UNIFIED CONCEPT OF NATURE

In the twentieth-century attempt to account for the physical world in a complete fashion, the quantum physics of the microcosm is complemented by a relativistic physics of the macrocosm. Of course, a genuine unification of these approaches has long been sought and still eludes us. Now the problem of infinity that disrupts quantum physical analyses has its counterpart in Einstein's general theory of relativity. It can be shown that the infinities in question in these two cases are identical. I am not merely suggesting that infinities of the *same type* arise in domains that are remote, domains at opposite extremes on the scale of magnitude. Rather, the point is that the appearance of such infinities signifies a *breakdown* in that scale, revealing the purportedly separate regimes to be one and the same. It follows that if the problem of infinity could be solved in a nonarbitrary, proper manner, the unification desired also would be achieved. Indeed, the *im*proper strategy of arbitrarily denying resurgent infinity so as to retain analytic continuity is one that preserves scale distinctions, thereby preventing anything but a superficial unification of microworld and macroworld.

Philosopher Milic Capek implied that Einstein's theories of relativity point toward a loss of analytic continuity, this result being manifested most clearly in the general theory. As Capek would put it, a "dynamization of space" is involved (1961, pp. 168–85). Classical analysis may be described as a procedure for reducing the dynamic to the static. This is the essence of mathematical derivation, as we have seen. Motions, actions and interactions, dynamic patterns that would otherwise express themselves as infinities, uncontrollable discontinuities, can be mathematically derived, expressed in static form, because the space in which they occur is itself simply continuous. Thus, we speak of transformations in space, but never the transformation *of* space, for space is classically pictured as an immutable, inert, three-dimensional container for *lower*-order dynamics.

Yet the transformation *of* space is precisely what the general theory of relativity entails. The theory meets the need to account for nonrectilinear, accelerative motion—identified with gravitational effects—in a relativistic manner. In so doing, Einstein found it necessary to depart from the standard Euclidean formulation and introduce the notion of *spatial curvature*. Gravitation is to be understood as the curvature *of* space, not as the product of forces acting *in* space, where space itself would remain impassive, wholly retaining its simply continuous character. Though the loss of simple continuity is suppressible as long as the curvature remains finite, when the role of curvature is fully played out, the consequence for analytic continuity becomes obvious. Solutions to the field equations for general relativity that predict *infinite* curvature indicate a *complete failure* of simple continuity.

Now the radical discreteness implicit in general relativity can be said to have a second aspect, for the theory additionally incorporates the idea that gravitational mass is *equivalent* to inertial mass. Again Capek's examination provides a helpful beginning. He observed that in pre-Einsteinian physics, inertial mass is interpreted as "the material substance itself [. . . the] core of matter . . . the substantial nucleus of matter," whereas gravitational mass can be viewed as matter's "action on space" (1961, pp. 178–79). Had Einstein completely succeeded in reducing inertia and gravitation to simply continuous field expression, they would have been made equivalent in the sense of being simply identical. The distinction between them would have been abolished. But in light of the irrepressible *discreteness* that develops in the general theory, perhaps we may reinterpret the equivalence relation as being more akin to the sort found in the *quantum* physical context, a Bohrian *complementarity* in which inertia and gravitation are indeed aspects of the same reality but not simply identical.

Have we not already discussed such a dual aspect relation in the context of quantum physical discreteness as such? From one perspective, matter is considered in and of itself, its *internal* effect on space being disregarded; from the other, matter per se is ignored, and its spatial manifestation takes the foreground (see fig. 4.1). Then we may associate the infinite gravitational mass of general relativity, that is, the infinite curvature, with the *spatial* infinity of the quantum domain. Here the climactic influence of discreteness is viewed as arising within continuity (fig. 4.1a). And the infinite inertial mass of the general theory may be identified with the *particulate* infinity of the microworld, for here the influence of discreteness is seen as imposed upon continuity from *without* (fig. 4.1b). According to the proposed reinterpretation of the principle of equivalence, the spatial and particulate forms of the infinite would be complementary aspects of the same underlying reality.

Figure 4.1

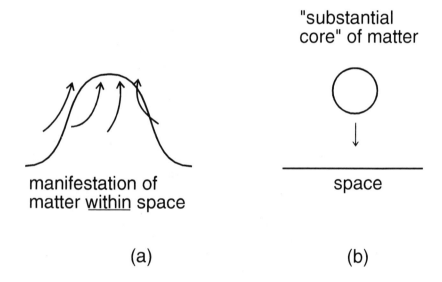

"substantial core" of matter

manifestation of matter <u>within</u> space

space

(a)   (b)

Schema for complementary influence modalities of matter/discreteness upon space/continuity: (a) spatial/gravitational mode—internal locus of discreteness; (b) particulate/inertial mode—external locus of discreteness.

However, the infinities of general relativity deal with large-scale effects—astrophysical events such as the gravitational collapse of massive stars—do they not? Must we not distinguish microcosmic discontinuity from that which manifests itself at the other end of the scale of magnitude? In fact the contrary is indicated, because the appearance of "macrocosmic" discontinuity signifies a *destruction* of the linearly conceived, Euclid-based scale, as we shall see shortly.

Einstein himself resisted the radical consequences of his theory, for while he was introducing a fundamentally nonclassical idea, he did not wish to relinquish the analytic power of the classical approach. So he adopted essentially the same strategem as his quantum physicist colleagues and paid the same price of self-contradiction. Einstein presupposed that non-Euclidean effects are not found at the microlevel, that is, that spatial curvature, incipient discontinuity, vanishes in the small. Locally, Einsteinian space is homeomorphic to Euclidean space. By assuming Euclideanness-in-the-small, Einstein could write equations using a formalistically elaborated version of the conventional calculus, equations designed to reduce the basically nonclassical, non-Euclideanness to classical terms. This amounted to an axiomatic invocation of simple continuity in the face of discontinuity so as to transform away the latter. We may study the attendant self-contradiction from the standpoint of its progressive development.

Under Einstein's classical assumption about the microworld, mass and spatial volume each approach zero as a limit. What does the notion of magnitude consistent with this assumption lead us to expect when we proceed up the scale? Increases in mass should be associated with proportionate increases in volume. Bodies of considerable mass should occupy considerable space, and the condition of infinite massiveness would be obtained only upon realizing infinite volume.

The general theory of relativity actually predicts the opposite. When we advance "up" the scale of magnitude, the volume of space does not increase in proportion to mass. This is because mass is associated with the curvature or *de-Euclideanization* of space, which, in turn, implies a germinal loss of spatial continuity. We need to understand that mathematical continuity means infinite *spatial density* (which is inversely proportional to the density of matter *in* space): an indenumerable infinity of points lies between any two points (however close) in such a totally "filled in" space. To curve space is to lessen its density. This produces a relativistic effect of scale contraction.

The effect is small in the middle ranges of magnitude but becomes quite conspicuous in the presence of great massiveness. Ultimately, at the "scale extreme" of infinite massiveness where the classical requirement is of infinite volume, the volume of space is zero because the density of space is zero. Space has gone singular; a "hole" has appeared. Euclidean space is demolished, or, in the language of physicist David Bohm (1973), space has become "enfolded."

In view of the fact that the general theory of relativity ends in an utter inversion of the classical concept of scale, it is obviously inappropriate to characterize gravitational collapse as a "macrocosmic, as opposed to microcosmic" phenomenon. On the contrary, we must say that gravitational collapse constitutes a *nullification* of this distinction. Only the deeply ingrained habit of Euclidean thinking disposes us to regard the relativistic gravitational effect as "macrocosmic." It is so compellingly natural for us to consider extreme massiveness as a large-scale phenomenon that we lose sight of what the radical de-Euclideanization of space actually accomplishes: it returns us to the *microcosm of quantum physics*, the problematic domain allegedly avoided by Einstein!

The crisis besetting Einstein's theory—articulated by physicist Brandon Carter (1968), for instance, in the doubts he raised that the theory could survive its prediction of gravitational collapse—is that its end result implies a radically non-Euclidean microstructure that rudely contradicts its initial premise of an entirely Euclidean one.

What we are witnessing again is the futility of attempting to contain resurgent infinity. The fate of the analytic scheme dependent upon the impotence of infinity is sealed in the domain where infinity is revitalized. In this connection, researcher Don Reed observed: "the equations [predicting "black holes," that is, gravitational collapse] are. . . consistent with the conventional. . . (classical) conception of infinity. . . .[However,] *black hole 'infinity' is foreign to any current conceptions of infinity that we are familiar with, involving a higher level of reality 'beyond' space and time*" (1980, pp. 41–42).

The crux of Einstein's difficulty, then, was an inability to handle this "black hole 'infinity,'" the self-same irreducible infinity currently confronting the quantum physicists. Clearly, as long as efforts to deny the dramatically non-Euclidean character of modern physical phenomena continue, a properly unified conception of nature will be unattainable.

The physicist Arthur Eddington, for one, was quick enough to admit that the curvature cannot really be denied. He identified the microphysical curvature with the "scale uncertainty." His advice: disregard it in favor of the conventional flatspace approach to the quantum regime "so that there is no loss of rigour" (1946, p. 46). Eddington's trepidations are as understandable as Einstein's, for once fundamental curvature is accepted, the tool physicists have relied on for centuries—the standard calculus with its method of limits—loses its universal usefulness. His message then is this: it is better to ignore a microlevel non-Euclideanness that would require developing an entirely new approach, and to proceed as before, though certain "adjustments" do have to be made now to keep the formalism viable. Yet the fact is inescapable that the necessity for such "adjustments" is tantamount to methodological failure. And it is the choice to continue making them that keeps us from the unity so urgently required.

TOWARD A SECOND PRINCIPLE OF THE RELATIVITY OF THE INFINITE

In sum, the crisis in modern theoretical physics is a crisis concerning infinity. The infinities of the classical era are a weak, relative, "garden variety," for these, in fact, have already been actualized or finitized with respect to our analytic, perceptual/cognitive repertoire. But they are giving way to a tough new strain, one creating problems and paradoxes for us that appear insurmountable. We may resist, stubbornly cling to the established order of the finite. We may refuse to compromise. But in that case, infinity will not compromise; it will persistently assume an absolute, unmanageable form. Then is there any alternative but to accept the infinite?

Now, the quantum level, "black hole" infinity is certainly not relative in the first sense of the total relativity of the infinite, but if it were simply absolute, accepting it would mean passive capitulation. There would be no choice but to resign ourselves to the end of the scientific enterprise. Ultimately, we would be obliged to surrender our analytical capacities once and for all, having lost our hold on space, time, dimension.

On the other hand, what if the "black hole microspace" could be viewed as a *new* dimension instead of merely a non-dimension? I mentioned this possibility earlier. An *embryonic dimension* would be one that the analyst does not yet have at his or her disposal—an infinity not yet finitized with regard to the analyst's repertoire. An altogether different attitude of acceptance would be called for here. The infinite could be accepted in an *active* manner, as a challenge to be met, for an embryo is a promise of renewal.

Supporting the notion of nascent dimensionality is the idea of *dimensional generation [see chapter 5]*. This is not an analytic theory but what we may consider a "*trans*analytic meta-theory." Analytic theories share the assumption epitomized in the Euclidean philosophy of Immanuel Kant: space and *time*—the latter a concept admittedly neglected in the present exposition— are a priori forms of awareness, fixed and unqualifiable givens to which events, contingencies, dynamisms are reduced. From the classical standpoint, the dynamic must yield to the static. In contrast, the idea of dimensional generation makes dynamic process basic by infusing space and time themselves with historical attributes. A perspective on spatiotemporal dimensionality is suggested from which the classical, analytic view can be seen as but a phase in a *developmental* context possessing a *pre*classical origin and a *post*classical terminus. Beyond even that, the meta-theory provides what we expect it must, as a concept that makes ongoing process fundamental: there is a regeneration of analytic capacity at a more comprehensive level. Thus, the meta-theory is transanalytic.

As has been noted earlier, Muses's introduction of dimensional generation and my own speak to the issue. Also, the work of the mathematician Benoit Mandelbrot (1977) on "fractals" (related to fractional dimension) furnishes some important insights, as does the more recent attempt of Frescura and Hiley (1980) to mathematize Bohm's "implicate order."

I believe that this kind of effort not only lends credence to the proposed interpretation of the quantum "dimension" as embryonic but may constitute the first step toward bringing the "embryo" to term. *Actively* accepting the "built-in discreteness," the "irremediably" infinite, means modeling it intrinsically, embodying it without attempting to reduce it, as *analytic* modeling would. In the *trans*analytic approach, discontinuity is accepted by a natural generalization of algebraic and geometric structure that affords a deeper continuity, an impartible, therefore unanalyzable, unity. Then, when the meta-theory achieves its fullest expression, we see the intimate interplay of deeper continuity with continuity of the simple, differentiative sort, a dialectic through which particular analytic regimes are destroyed and new ones unfold. This transformative scheme is the one foreshadowed in my previous presentations of the "Moebius-Klein" meta-geometry [see chapters 1–3], and, if I am not misinterpreting his thinking, it is inherent in Charles Muses's (1977) algebraic concept of "hypernumbers," the hypernumber w playing an especially significant role.

It is important to realize that such radically nonlinear meta-modeling of dimensional embryogenesis will be no mere exercise in armchair abstractionism—not if it is inherently valid, not if its profound methodological implications are pursued to their conclusion with intuitive honesty. An *ad hoc* description, one that is arbitrary and self-serving, will bear little fruit. But a description that faithfully adheres to the "underlying mathematical reality," as Muses (citing Godfrey Hardy) termed it (1977, p. 72), will have concrete creative consequences. In other words, to describe change in an intuitively coherent manner is constructively to *foster* change.

Here is our intimation of a second, more general principle of the relativity of the infinite. For, while the first principle of the *completely* relative infinity has presently failed us, I am proposing that we can deal with the infinite in its stronger manifestation, "finitize" it, so to speak, divest it of its absolute character. But not by beating it into submission, not by establishing supremacy over it from an entrenched position in our home territory. We must meet it halfway, that is, be willing to relinquish the *established* order of finitude, admit elements of potentiality that will transmute and expand what was already actualized. The essential idea is enunciated by Francisco J. Varela in his discussion of paradoxes: "Instead of finding ad hoc means of *avoiding* their appearance (as in Russell's theory of types) we let them appear freely" (1975, p. 21). To illustrate, Varela draws from the history of arithmetic. Mathematicians were able to extend the real number domain to that of the complex numbers by accepting the idea that a number (namely, the "imaginary" number i) may be both positive and negative. "Again, rather than avoid the antinomy, by confronting it, a new domain emerges" (1975, p. 22).

5

A Neo-Intuitive Proposal for
Kaluza-Klein Unification (1988)

If we had a consistent way of looking at the significance of quantum mechanics intuitively and imaginatively, this would help us to avoid allowing our attention to be...diverted from the novel implications of quantum interconnectedness. Such a way of looking might also help to indicate new directions of theoretical development, involving new concepts and new ways of understanding the basic nature of matter, space, time, etc.
D. J. Bohm and B. J. Hiley *On the Intuitive Understanding of Nonlocality as Implied by Quantum Theory*

1. INTRODUCTION

A fundamental aim of contemporary theoretical physics is to unify the forces of nature in a single account. Recently, attention has focused upon a revitalized version of the so-called Kaluza-Klein approach to unification (Scherk and Schwarz 1975; Cremmer and Scherk 1976). In his original paper, Kaluza (1921) suggested reconciling Einsteinian gravitation with the electromagnetic force by expressing the latter in terms of a fifth dimension added to the four (three spacelike, one timelike) that constitute the known universe. Subsequently, Klein (1926) showed that it is possible to write Shroedinger's wave equation in five independent coordinates, thereby demonstrating the basic compatibility of Kaluza's proposal with quantum mechanics. Regarding the question of how a fifth dimension could be accommodated in a universe apparently limited to only four, it was supposed that the additional dimension could be compactly curved, compressed at a scale of 10^{-32} cm.

In the years following the conjectures of Kaluza and Klein, technological advances permitted the construction of particle accelerators by means of which two new fundamental forces could be studied: the strong and weak interactions. At the subatomic scale on which these forces operate, gravitation plays a negligible role. Therefore, in devising gauge theories to unify short-range forces as members of the same abstract internal symmetry group, physicists could disregard the interrelated issues of space-time symmetry and dimensionality more directly relevant in the context of general relativity. During this phase of the quest for unification, development of a Kaluza-Klein theory was not a high priority. But by the late 1970s, weak and electromagnetic forces successfully had been unified (by Weinberg and Salam), and an effective theory of the strong interactions formulated (quantum chromodynamics); moreover, the prospects seemed good for a grand unification encompassing all three forces. At this time, the deferred question of quantum gravity reasserted itself, and this renewed interest in a Kaluza-Klein program, which was now to incorporate the gauge-theoretic idea of "spontaneous symmetry-breaking."

In this exposition, I will explore at a fundamental level the conceptual implications of revitalized Kaluza-Klein theory. Here it will become clear that unification cannot adequately be addressed merely as a theoretical problem, that certain critical *philosophical* questions arise that need to be confronted in a full and effective treatment of the matter.

To begin, I will demonstrate that there is a sense in which the Kaluza-Klein approach appears to undermine the intuitive foundations of mathematical physics. A discussion will follow of how this implicit consequence has been repressed at a substantial cost (section 2). But I shall advocate neither discarding "the baby with the bathwater" nor remaining in the "dirty tub." Instead of denying the Kaluza-Klein potential for unification so as to uphold the purity of classical intuition, or continuing to pay an unacceptable price for maintaining Kaluza-Klein within the classical framework, I shall reformulate the Kaluza-Klein approach, casting it within a new intuitive context seen to contain the old as a limit case (section 3). Interestingly, application of this strategy to the problem of cosmogony leads to theoretical results that accord with the overall ten-dimensionality of the recently favored super-string species of Kaluza-Klein. Moreover, at a finer level of analysis, new cosmogonic features are predicted regarding the particular pattern of "dimensional compaction," features that should be difficult to anticipate within the current framework (section 4). Nevertheless, it would be misleading to emphasize the quantitative consequences of the proposed strategy, since this would tend to obscure the fact that these "neo-intuitive" findings arise, and must be interpreted, within a conceptual context that is *qualitatively* different from the traditional one. The primary objectives of this paper are to show the need for such a context and to communicate its essential characteristics. In the concluding section (5), the relationship between neo-intuitive unification and David Bohm's "quantum interconnectedness" will become evident.

2. DOES SPONTANEOUS SYMMETRY-BREAKING IMPLY A CONCRETIZATION OF MATHEMATICAL PHYSICS?

2.1 Apparent Concretization of Mathematical Physics via Symmetry-Breaking

A primary strategy of twentieth-century theoretical physics is to describe the laws of nature in terms of mathematical symmetry. In general, a symmetry is defined when some characteristic of a body or system remains the same despite the fact that a change has been introduced. For example, if a sphere is transformed by rotating it through any angle about its center, its appearance will not change. The sphere therefore can be said to be symmetric under the operation of rotation. This simple notion of symmetry is generalized in group theory, where a variety of mathematical systems can be classified in terms of the groups of transformations under which they remain invariant. Applying the approach to theoretical physics, the laws of physical interaction are described as abstract symmetries. Thus, in the framework of special relativity, we can say that the form in which electromagnetic interactions occur (as described by Maxwell's equations) remains invariant under global (Lorentzian) transformations of space-time coordinates, and in general relativity, the equations for gravitational interaction are invariant under local (Riemannian) transformations of coordinates. The technique has been especially emphasized in quantum field theory, where invariance in the form of physical interactions has been studied under transformations of "internal space" (here properties such as charge, spin, and "color" are examined via transformations of "Hilbert space" *[see chapter 4]*). Now, if the laws of physics effectively can be expressed in terms of abstract symmetry relations, would it not be possible to define more general symmetry groups under which two or more kinds of physical interactions could be subsumed as subgroups? Such an extended application of the idea of symmetry constitutes the basic rationale for unification. The first contemporary example was mentioned above: the electroweak unification achieved independently by Weinberg and Salem in 1967–68. But with this accomplishment, the potential for a fundamental change in philosophical orientation was subtly introduced.

Until the time of the Weinberg-Salam breakthrough, the mathematical purity of theoretical physics had never been exposed to the prospect of compromise. Understood most essentially, this "purity" depends on preserving a distinction whose lineage can be traced as far back as Kant and Descartes and, ultimately, Plato: the distinction between the "normative" and the "empirical" (Hooker 1982), that is, between form and content, abstract reason and concrete fact. The latter is associated with mutability, dynamic process; the former with changeless structural relations axiomatically presupposed. With this dichotomy built into it at the deepest level, the *modus operandi* of mathematical physics has been to express one member of the pair in terms of the other: concrete process is re-presented as abstract form. Indeed, this is but another way of characterizing the symmetry-group approach, wherein the concrete dynamics of nature are accounted for in terms of nature's invariant forms (i.e., its structural laws).

Now, in raising the question of unification, emphasis is shifted from concern about deriving concretely observable differences from abstract uniformity, to concern about differences among the abstractions themselves. Given the fact that physical processes simply do not lend themselves to expression in fewer than four distinct forms, how could these four symmetries be rendered symmetric with respect to *each other*, reduced to a single symmetry? The answer proposed by Weinberg and Salam was that these symmetries *were* symmetric with respect to each other in an early phase of cosmological development but that this all-embracing primordial symmetry subsequently was broken spontaneously. This suggestion, now widely accepted by theoretical physicists, seems as if it could imply an extraordinary reversal of roles between process and structure.

In the long-sanctioned approach of attempting to express dynamic occurrences in terms of abstract symmetries, theoretical physicists, in effect, were seeking to relegate concrete change to secondary status, render it an epiphenomenon of changeless structure. But in the concept of 'spontaneous symmetry-breaking' required for unification, physics is wedded to cosmogony, and in consummating the marriage, cosmological time appears to emerge as the measure of formal structure! Does this not amount to a concretization of symmetry relations that previously were regarded as pure abstractions, analytical devices for describing nature that were themselves outside of nature, exempt from its dynamics? Now it is no longer merely the *contents* of nature that are to be viewed as changing in accordance with nature's invariant form; it seems we must view the form itself as being subject to historical process.

This apparent challenge to epistemological tradition can be clarified further by stating it geometrically, in terms of dimensions, rather than symmetries. The geometric equivalent of the classical assertion that nature remains invariant in form is that spatiotemporal dimensionality is invariant. Indeed, this was a prime dictum of Kant, who held that particular perceptions are contingent, always subject to change, but that all perceptual awareness is organized in terms of an immutable intuition of space and time. In the words of Fuller and McMurrin, Kant took the position that "no matter what our sense-experience was like, it would necessarily be smeared over *space* and drawn out in *time*" (1957, pt. 2, p. 220). Implied here is the categorical distinction between the empirical and the normative, ever-changing concrete fact and unshakable reason. We observe fact *by means* of reason; we do not observe reason. For Kant, space and time are in the latter category. They constitute the analytical framework for experiencing change, a framework that is itself impervious to change. Is the dimensionality of space and time, its immediately intuited 3 + 1 character, a fact? Not at all. According to Kant, it is given a priori, as an article of pure reason, and thus is not susceptible to alteration. Formulated in geometric terms that would appear to challenge this classical intuition of space and time, the notion of cosmogonic symmetry-breaking takes the form of a *Kaluza-Klein theory*.

Until the recent revival of Kaluza-Klein noted in the introductory section, the Kantian doctrine had not been seriously questioned. In the relativity theory, space and time were mathematically combined to yield a "four-dimensional" continuum, but in actuality, their independence had not simply been abolished, and theorists like Weyl (1922) insisted that we continue to view space-time as a 3 + 1-dimensional structure.

Importantly, relativistic space-time was as much an abstraction as independently construed space and time had been; therefore, no direct contradiction of Kant's intuition was indicated.[1] Nor was that intuition compromised in quantum electrodynamics, where n-dimensional "Hilbert spaces" were introduced largely for analytical purposes but the 3 + 1-dimensional structure of macrophysical space-time was not questioned. As for the original Kaluza-Klein unification attempt, there was no suggestion that a fifth dimension had been invoked for any other reason but mathematical convenience. Then came the Weinberg-Salam unification, its key idea of spontaneous symmetry-breaking leading to the new, more intimate relationship with cosmogony. It is in this novel context that the Kaluza-Klein program of added dimensionality has been revitalized, and in the process, the concept of 'dimension' seems to have undergone an unprecedented reversal of status: it seems to have become "empiricized."

According to the basic Kaluza-Klein interpretation of cosmogony, the primordial symmetry condition of the universe is represented by a compact, multidimensional manifold, all dimensions being "real" ("regarded as true, physical dimensions"; Witten 1981) and all co-existing at the same microcosmic scale (close to the Planck length). The event of symmetry-breaking is identified with what may be called "dimensional bifurcation": a subset of dimensions expanded relative to the remaining dimensions and this broke the initial equilibrium. The expanded, 3 + 1-dimensional universe known to us today—far from being given a priori as an article of pure reason—is the result of the cosmogonic process of dimensional transformation. Thus, the apparent consequence of the renewed Kaluza-Klein approach is that spatiotemporal dimensionality is no longer to be regarded strictly as a changeless framework for change. Dimensionality itself is seemingly thrown into the arena of concrete change, thereby mitigating the absolute distinction that had been drawn between structure and process, reason and fact.

But has such a radical step actually been taken in the *mathematical foundations* of physics?

2.2 The Avoidance of Dimensional Symmetry-Breaking and Its Consequences

Although Kaluza-Klein cosmology may appear to suggest dimensional transformation, near the beginning of this century, when mathematical intuitionists were laying the topological foundation stones of contemporary dimension theory, the Kantian dictum of dimensional *invariance* was tacitly incorporated. To begin with, in 1911 Brouwer was able to prove that dimensions are not strictly homeomorphic with respect to each other: no operation is permissible by means of which one dimension can be put into 1:1 correspondence with another and, also, be continuously mapped into the other. What still was needed for a clearcut demonstration of the topological invariance of dimension was the positive identification of a simple topological property responsible for the nonhomeomorphism. This came in 1912 with the intuitive reasoning of Poincaré. In Poincaré's scheme of classification, a given continuum could unambiguously be assigned a fixed integer value unique to it, its dimension number, said number depending on the number of continuous cuts required to divide it. Subsequently, Poincaré's definition was precisely formulated by Brouwer and further refined by Urysohn and Menger (see Hurewicz and Wallman 1941, pp. 3–5). The effect of all this was to guarantee that dimensions themselves indeed were invariant, not subject to transformation. They were to be accepted "ready-made," assumed to provide the uniform context within which all other geometric transformations were to be described.

To grasp more essentially the classical conception that was at stake, we consider Poincaré's intuitive definition of dimensionality in a more detailed manner. In his initial attempt to define dimension unambiguously, Poincaré reasoned as follows:

> If to divide a continuum C, cuts which form one or several continua of one dimension suffice, we shall say that C is a continuum of *two* dimensions; if cuts which form one or several continua of at most two dimensions suffice, we shall say that C is a continuum of *three* dimensions; and so on.
>
> To justify this definition it is necessary to see whether it is in this way that geometers introduce the notion of three dimensions at the beginning of their works. . . Usually they begin by defining surfaces as the boundaries of solids . . ., lines as the boundaries of surfaces, points as the boundaries of lines, and they state that the same procedure can not be carried further.
>
> This is just the idea given above: to divide space, cuts that are called surfaces are necessary; to divide surfaces, cuts that are called lines are necessary; to divide lines, cuts that are called points are necessary; we can go no further and a point can not

be divided, a point not being a continuum. Then lines, which can be divided by cuts which are not continua, will be continua of one dimension; surfaces, which can be divided by continuous cuts of one dimension, will be continua of two dimensions; and finally space, which can be divided by continuous cuts of two dimensions, will be a continuum of three dimensions. (Quoted in Hurewicz and Wallman 1941, p. 3)

A deeper intuition is implicit in the procedure for assigning dimension number intuited by Poincaré. In Poincaré's discussion of "continua," one of the most fundamental ideas of classical philosophy is presupposed: that of *res extensa*. Space, in its essence—the space of Plato, Euclid, Descartes, and Kant—is continuity; and continuity, as given a priori to classical reason, entails *extendedness*. Consider, as an illustration, the one-dimensional space represented by a line segment. The first dictate of intuition clearly is that this line has finite extension. Then it must be continuous; it can possess no holes or gaps in it, since, if the point-elements composing it were not densely packed, we would have not a line at all but a collection of unextended points. The quality of being extended implies the infinite density of constituent point-elements.

Yet, at the same time, reflection discloses that the classical continuum possesses a property that prompted Muses (1968) to refer to it as a "*dis*continuum." For the absence of gaps not only holds space together but also permits it to be indefinitely *divided*. Without a hole in space to interrupt the process, there is no obstacle to the endless partitioning of space into smaller and smaller segments (note the association of continuity with divisibility appearing in Poincaré's assertion that "a point can not be divided, a point not being a continuum"). As a consequence, though the points constituting the linear continuum indeed are densely packed, they are distinctly set apart from one another. However closely juxtaposed any two points may be, a differentiating boundary permitting further division of the line always exists. As Capek (1961) put it in his critique of the classical notion of space, "no matter how minute a spatial interval may be, it must always be an *interval* separating two points, each of which is *external* to the other" (p. 19). The infinite divisibility of the extensive continuum also clearly implies that the constituents are themselves unextended. Consequently, the point elements of the line can have no internal properties, no structure of their own.

Generalizing this understanding of the line to higher-dimensional entities, we may summarize the classical intuition of space in the following way: A space of any dimension is an extended, infinitely divisible continuum whose densely assembled elements are sharply set off from one another and are devoid of intrinsic structure.

To see how the intuition of continuity is related to the idea of symmetry, I reiterate the general definition of the latter: The study of symmetry is concerned with the properties of an entity that remain invariant when some transformation is introduced. Felix Klein employed this approach in the Erlangen Program (1872) by which he classified the new forms of geometry that had come to be investigated in the nineteenth century. Euclidean geometry, for example, is definable as the set of postulates that remain invariant under rigid (distance-preserving) transformations, but no others. In contrast, the properties of projective geometry (such as that expressed in Desargues's theorem) remain invariant under central projection, an operation that creates the effect of *changing* metrical relations (e.g., transforming a circle into an ellipse). The least restrictive, most general form of geometry is topology; its only demand is that, whatever transformation we introduce, it be a *continuous* transformation. Thus, in topology, a circle can be transformed not only into an ellipse but also into any other geometric object, provided this is done without disrupting the object (e.g., by tearing it or punching a hole in it) so as to cause "the loss of identity of its points" (Karush 1962, p. 272). This amounts to saying that, in topology, just one invariance requirement need be met: that of space itself. Topological invariance requires only that the classical intuition of the continuum be preserved. Naturally, since the transformation groups defining other geometries are subgroups of the more general topological group, they too must conserve the continuity of space. Thus, extensive continuity may be regarded as the fundamental symmetry of geometry.

Does extant Kaluza-Klein theory entail a breaking of geometry's fundamental symmetry? In the simplified schema for the standard formalism given in figure 5.1, no transformation of dimensionality as such is seen to occur. An initially "compact" two-dimensional manifold is depicted as expanding to observability along its horizontal axis, its vertical dimension remaining "microcosmically scaled." This differential expansion breaks the

Figure 5.1

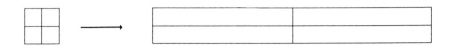

Schema for standard formulation of Kaluza-Klein cosmogony

perfect symmetry of the continuum, but underlying *spatial* symmetry (as defined above) is not broken; continuity is retained, the transformation being merely topological. By analogy, we may imagine a primordial manifold of ten dimensions (to use the "superstring" interpretation of cosmogony) being topologically transformed so as to produce our presently observable 3 + 1-dimensional universe. On this account, intrinsic dimensional change is rendered apparent. That is, the event of "dimensional bifurcation" implies no *intrinsic* reduction in the dimensionality of the original manifold. The manifold remains ten dimensional, the mere *appearance* of 3 + 1-dimensionality being created by changes in the global structure of space.

Nevertheless, while the Kaluza-Klein formulation preserves the classical concept of dimensional invariance, it does so in a manner that actually undermines the intuitive foundations of classicism established by Descartes and Kant. The symmetry of space, its extensive continuity, is maintained by the epistemologically costly procedure of resorting to a higher order of abstraction.

With continuity serving as their first principle, *both* classical and modern unification theory preclude the possibility of inherent change. Since the continuum itself is changeless, it can provide no intrinsic directive for the concrete dynamics of nature. Its relation to dynamic reality is expressed through its *ex*trinsic structure, its topology. The topological properties of space are the structures that mediate between space's internal[2] changelessness and observed physical processes; empirically determined dynamics are representable in topological form.

Now, in classical physics, the gap between the mathematical representation of physical reality and dynamic reality itself could be kept to a minimum. In the classical equations of motion, for example, changes observed in nature were directly expressed.

Indeed, this capacity to bring concrete change and formal description into close alignment was of critical importance in the enterprise of classical physics. What permitted this alignment to be achieved? The intuitive certainty of space and time as the context for experience, the unchanging *form* of experience that framed all its concrete contents in an immediate manner. By contrast, modern unification theory permits a measure of contact to be lost between form and fact. They are removed from each other by introducing a second order of abstraction in relation to which the old order is treated concretely. The space and time of immediate experience are no longer taken as composing the fundamental abstraction. Instead they are regarded as themselves merely extrinsic, topological manifestations of a higher-dimensional continuum unknowable by direct intuition. Because the "primordial continuum" of Kaluza-Klein theory does not constitute the directly intuited framework for our experiences, no topological expression of it can be put into direct correspondence with experience. At what point and in what specific way does the opening of this gap become a problem in the standard account of symmetry-breaking?

In the supposed transition that was made from the primordial state of symmetry to the topologically less symmetric present condition, two symmetry subgroups are assumed to have been produced: that associated with the 3 + 1-dimensional observable universe, and the one corresponding to the submanifold that has remained compact. It is the compactness of the latter, its submicroscopic concealment, that presumably creates the mistaken impression that the 3 + 1-dimensional universe is what constitutes the fundamental symmetry; mathematical descriptions of the forces of nature have been so different from each other because they have been written in terms of that symmetry, thereby neglecting the hidden symmetry. The essential task of unification is seen as recovering the hidden symmetry. Unity is presumed to be attainable by describing the observable and compact manifolds as the subsymmetries of an overall symmetry topologically expressed in higher-dimensional space. The problem is that since this topological structure in principle cannot be directly aligned with what we experience, it must necessarily be determined by an indirect, less-convincingly grounded procedure.

When only two forms of interaction are to be unified (as in electroweak unification), ambiguity is minimized. The ideal case is that in which there is no latitude for variation in the topological characteristics of symmetry components, thus, only one way to recover the hidden symmetry. However, as the number of forces to be unified increases above two, necessarily it will become possible to satisfy the n-dimensional field equations with more than one topological solution; and when four forms of interaction are to be encompassed, the array of topological possibilities is staggering. Naturally, practical methods exist that are designed expressly for the purpose of reducing ambiguity, narrowing the field of possibilities. Yet the essential fact remains that the standard approach, entailing as it does an order of abstraction one step removed from that aligned with concretely observable reality, provides no means for directly ascertaining the topological structure presumed to have arisen with symmetry-breaking. Therefore, when this approach is employed, efforts to recover hidden symmetry eventually become cumbersome and work must proceed in the dark.

The reader familiar with issues in the foundations of mathematical physics may recognize my "intuitionist" critique of Kaluza-Klein formalism as one equally applicable to quantum mechanics (QM) in general. In the standard, Copenhagen interpretation of QM, no bones are made about the fact that its Hilbert "space" is but a mathematical fiction devised to yield "pragmatic results" in the absence of an ability to represent the microphysical domain in an intuitively grounded, realistic manner. And while opponents of the Copenhagen view have argued for a "realistic" construal of the quantum domain, the "reality" most describe arises as a mere artifact of the *formalism [see chapter 12]*; thus it must be regarded as being but nominally real. It is in essentially the same sense that the standard Kaluza-Klein account of cosmogony is "realistic." Like the subspaces of QM's Hilbert space, the subsymmetries of Kaluza-Klein do not reflect concrete reality as much as they reflect the need to uphold abstract continuity. It is therefore not surprising that the shortcomings of Kaluza-Klein pointed out above are closely related to those observed of QM proper, by the minority group of theorists who have sought a form of realism that is more than merely nominal (e.g., Bohm 1980; Stapp 1979; Josephson 1987). But this does not mean that formalism should be abandoned in favor of *traditional* realism or intuitionism.

The mere renunciation of formalism in advocacy of classical intuition cannot address the problem of unification. I suggest that the basic difficulty with the abstractionist approach to unification lies not in its questioning of classical intuition but in its failure to question it deeply enough. Though the Kaluza-Klein program implies that unification indeed cannot be achieved if we restrict ourselves to the Kantian intuition of space and time, instead of raising genuine doubt about that intuition by opening continuity itself to question, we have seen that Kaluza-Klein theory *maintains* the continuity abstraction in a higher-order form. Thus, both the need to question classical intuition and the intention of upholding it are implicitly expressed. By indicating that the old intuition of space and time should not be taken a priori after all, that it actually resulted from a concrete cosmogonic process, the question is raised; by at once assuming that the space-time continuum of immediate experience is but a topological manifestation of a more abstract order of changeless continuity, the question is cancelled. Having gotten an idea of the price the abstractionist pays for cancelling his own question before it earnestly can be asked, we may now begin to ponder the alternative proposed in this paper:

Suppose the question is not cancelled. Suppose a *genuine* concretization of mathematical physics is entertained. Instead of sidestepping the loss of Kantian symmetry, of intrinsic continuity, by rendering the 3 + 1-dimensional continuum a merely extrinsic manifestation of a more abstract continuum, suppose we hold our ground at the first level of abstraction and *allow* the classical intuition of space and time to be challenged. What would this mean? Could mathematical physics, which has always given primacy to symmetry, survive such a development?

Of course, to question the principle of symmetry is to venture beyond it. But rather than merely forsaking spatial symmetry for some notion of nonsymmetry that is foreign to it, what if nonsymmetry could be given *internal* expression? Then space would be dynamized, seen to "change from within itself," as it were. Would not this concretely self-transforming space provide an intrinsic directive for the details of cosmogony? In the absence of a gap between levels of abstraction, there would be no free-floating topological structures requiring indirect empirical determination, thus there would be no "free parameters." All structures would appear as immediately intuited concomitants of the transformation process, giving a full and natural unification of the forms of interaction manifest in nature.

3. SYNSYMMETRY AND THE GENERALIZED MOEBIUS CONCEPT: A NEO-INTUITIVE PRINCIPLE OF DIMENSIONAL GENERATION

3.1 Symmetry-Breaking as a Self-Transformation of Space: A Provisional Account

By introducing the possibility of *internalizing* the breaking of symmetry, we are asking what a new, more dynamic intuition of space would involve. Can we construct some model, develop some conceptual scheme in which the transformation of dimensionality is intuited as a concrete process of *self*-transformation?

As noted earlier, the classical concept of continuous topological transformation presupposes that the elements of space retain their well-defined identity. In Poincaré's formulation of topological invariance, the sharply defined, structureless elements are the "boundaries" or "continuous cuts" (points in a line, lines in a plane, etc.) that serve to distinguish one dimension from another in an unambiguous manner. If spatial continuity somehow were breached, boundary elements apparently would lose their sharp definition and topological invariance, the principle prohibiting spontaneous transformation of dimensionality, would be compromised. Thus, it seems the idea of self-transforming space requires us to entertain the prospect of the transformation of the identity elements of space themselves. While not much sustained attention has been given to this prospect, a few relevant speculations have been offered.

In the context of asserting that "the concepts of spatial and temporal continuity are hardly adequate tools for dealing with the microphysical reality," Capek (1961, p. 238) cited Menger's (1940) attempt to construct a "topology without points." Of course, it was not Menger's intention simply to drop the idea of space-time, so it was necessary for him to include *some* form of distinction. His dilemma led him to suggest a topology of "lumps" (instead of sharply defined points). Capek quoted Menger's own misgivings in this regard:

> by a lump, we mean something with a well defined boundary. But well defined boundaries are themselves the results of limiting processes. . .Thus, instead of lumps, we might use at the start something still more vague—something perhaps which has various degrees of density or at least admits a gradual transition to its complement. Such a theory might be of use for wave mechanics. (Menger, quoted in Capek 1961, pp. 237–38)

The "topology" envisioned by Menger certainly runs counter to classical intuition: A "space" without well-defined boundaries, one not possessing uniformly infinite density, thus not infinitely divisible either, not reducible to structureless identity elements. A hint of the internally dynamic character of such an entity is given in Menger's suggestion that it should admit "a gradual transition to its complement." In several other contexts, I have worked with what seems to be a closely related idea. The conceptual model I have proposed [i.e., the Moebius model], adapted for the problem of symmetry-breaking, may help bring Menger's insight to fuller development.

As a means of gaining intuitive access to the nonclassical notion of internalized symmetry-breaking, I make use of a familiar structure from classical topology: the Moebius strip (fig. 5.2b). To identify the transformational properties of this half-twisted surface, we compare it with a cylindrical ring, its untwisted counterpart (fig. 5.2a).

Figure 5.2

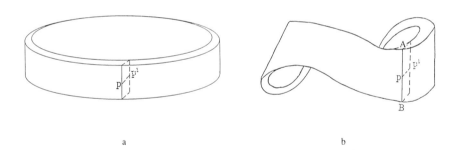

a

b

Cylindrical ring (a) and surface of Moebius (b)

The cylindrical ring is classified as a two-sided surface. Of course, conventionally, sidedness is interpreted not as a dimensionally intrinsic (i.e., local) property but as a topological one arising from the way the surface is embedded in three-space.[2] That is to say: As an intrinsically two-dimensional geometric entity, the surface is infinitely thin and cannot possess two sides, but by virtue of the fact that it is embedded in three-space, it will have finite thickness. Thus, two points, P and P', situated in diametrically opposite sides of the cylindrical surface, owe their distinct existence not to the intrinsic character of the surface but to the fact that the surface is extended extrinsically in an additional dimension (denoted in fig. 5.2 by interior line *PP'*).

Naturally, the Moebius surface (fig. 5.2b) also appears extended in the direction perpendicular to itself. Therefore, at any transverse section of the Moebius (illustrated by line *AB*), points in opposite sides (constituting bounding elements of an interior perpendicular line) are sharply definable; an unambiguous distinction exists between them, as in the case of the cylindrical ring. But while the Moebius model is two-sided with respect to its transverse axis, when its longitudinal extension is taken into account, it must be regarded as a *one*-sided surface. We may think of expansion along the longitudinal axis as an operation whereby opposite sides "dissolve" into each other to form but a single side. It is this interesting property that permits the Moebius to serve as a model of internal symmetry-breaking.

In the analogy, classical continuity is seen to prevail only in the two-sided limit case shown by cross-section *AB*, where points *P* and *P'* are well-defined elements of the interior line. Here *PP'* possesses infinite density. With expansion from this limit to the longitudinal condition of one-sidedness, *P* and *P'* are imagined to merge progressively, losing their distinct identity; the density of line *PP'* thus is viewed as diminishing. When the expansion is complete (the full length of the Moebius being encompassed), points *P* and *P'* are seen as entirely identified. Points that once were sharply differentiated now have ceased to exist in separation, and the density of the interior continuum has been reduced to zero. Moebius expansion therefore appears to provide a simple model for expressing Menger's intuition of a "topology without points." The Moebius operation symbolizes a process whereby space "breaks its own symmetry."

However, the foregoing account is but a first approximation, one that actually gives a truncated view of symmetry transformation, as will become evident when a more fully articulated version is developed in subsequent sections. In concluding the present section, let us note the obvious limitation of the Moebius analogy when evaluated in traditional terms.

Conventionally, the one-sidedness of the Moebius is treated in the same way as the two-sidedness of the cylindrical ring: it is regarded as a property resulting strictly from the manner in which a structure is embedded in space, rather than as being dimensionally intrinsic. Points P and P' become identified merely with respect to the *global* structure of space. The locally symmetric character of the continuum, its infinite density, never is actually affected. In upholding the symmetry-preserving intuition of Poincaré, dimensional invariance is upheld, dimensions remaining categorically distinct entities not susceptible to transformation. Here the topological dimension number of any mathematical structure must be a fixed integer value, the Moebius structure being no exception. Accordingly, its dimensional status is regarded as identical to that of the cylindrical ring: it is a strictly two-dimensional entity embedded in a three-dimensional continuum. In the articulation of the Moebius model to be carried out in sections 3.4–3.6, we shall see the sense in which the classical assessment of the Moebius strip is correct, and that in which it is limited.

3.2 From Symmetry-Breaking to Dimensional Generation

We may begin to understand the deeper limitations of the provisional model of "internal symmetry-breaking" by considering the general question of whether the very idea of symmetry-*breaking*, is, after all, wholly compatible with an attempt to internalize it. The description of cosmogony as a process of symmetry-breaking carries with it an important implication thus far not directly considered.

To speak of "breaking spatial symmetry" is to suggest a negation of symmetry that certainly would appear to be at odds with classical intuition. Yet does not this very assertion also presuppose an initial state in which symmetry was *un*broken? If such a state of perfect symmetry existed, its subsequent disruption would have to have been instigated by an agency external to it, for a purely symmetric space could not break its own symmetry. Here we well might wonder whether any nonsymmetric incursion from outside could breach an original symmetry at its *core*, that is, whether externally induced change could bring about *intrinsic* change. Of course, for classical intuition, such change is not intended. It is underlying *changelessness* that is mandated by the classical assumption that dimensional transformations wrought by "symmetry-breaking" are "merely topological."

In developing a new intuition of space, we cannot put symmetry first by speaking of "symmetry-breaking" but must view symmetry itself as arising from a more primordial, presymmetric condition, one in which symmetry is *already* "broken," as it were. Only when the nonsymmetric aspect is thus formulated in positive terms (rather than as a mere negation of a pre-existent symmetry) can it be internalized properly, enabling us meaningfully to speak of a self-transforming symmetry process.

This leads to a view of cosmogony quite different from the standard one. The need for a new intuition requires us to alter our conception of what constitutes primordiality. With symmetry losing its primordial status, the basic problem of cosmogony changes from that of explaining symmetry-breaking to that of understanding how symmetry is created in the first place. The *making* of spatial symmetry, the generation of dimensionality, becomes cosmogony's foremost concern.

In a speculative paper on "dimensional generation," Muses (1975b) attempted to carry intuition beyond the topological concept of dimension. His primary notion is that of the "fractional dimension,"[3] illustrated by the idea of a "partially generated line" (p. 17). Consider the relationship between the point and the line. While classical intuition demands that we assume the line to be infinitely rich in points, Muses would have us imagine this condition of infinite density being reached only after the line has been *generated* from the point. We are to intuit intermediary stages of production wherein the point density of the line is *less than* infinite, though greater than one. In introducing the idea of a fractionally generated line (not to be confused with the fully generated line segment, which *is* infinitely dense, however short), Muses noted "that the concept of point density bridges the otherwise separated and un-unified concepts of point and line" (Muses 1975b, p. 17).

How in particular is dimension $n + 1$ produced from n? By what process is point density increased to create a line? According to Muses, the simplest and most parsimonious solution is an evolving sine-wave function. The wave function, $y = h \sin (2 \pi t/\lambda)$, is given, where t is a space interval, h is wave amplitude and λ is wavelength. The function is diagramatically expressed in figure 5.3 (which simplifies Muses's [1975b, p. 18] original diagram). Muses commented that the "only 'real' or manifest portions of the dimension are the points on the wave axis at half wave-length

Figure 5.3

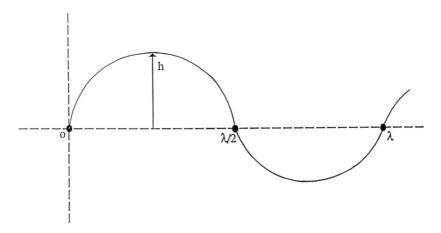

Dimensional wave. After Muses, 1975b

or $\lambda/2$" (p. 18). Now we are to picture the wave form changing so that "D approaches unity, i.e. . . . a part-line becomes a line. As this happens, the frequency of the wave markedly increases and its amplitude decreases till at $D = 1$ the wave-length is zero (i.e. infinite frequency) as well as the amplitude" (p. 18). In other words, since point density is directly proportional to wave frequency, an increase in the latter "fills in" the line.

However, while the general notion of dimensional development constitutes a challenge to the classical intuition of spatial continuity, Muses's particular formulation of it actually presupposes continuity. Muses's evolving sine wave is a *continuous transformation*. To posit such change is, at the same time, to posit the more fundamental invariance of the context in which it occurs. The transformation of the wave form by means of which the line is "generated" assumes the topological invariance of the two-dimensional plane of analysis. On Poincaré's intuition, such a continuum is *defined* by the fact that its boundary element is the well-formed line, an entity that, when taken as a continuum in its own right, would be infinitely dense. Thus, although the sine-wave analysis is intended to represent linear conditions of less than infinite density, the density of the "part-line," in fact, is never less than infinite. In using the two-dimensional plane of analysis to portray "density operations" on the line, the "part-line," in effect, is already filled in, already infinitely dense. Equivalently put, density variations in the line are not intrinsic but merely result from the topological treatment the line is given in invariant two-space.

The limitation of Muses's approach may be understood from another perspective, by considering the mistake of assuming that "a partially generated line should contain . . .some finite number of points" (Muses 1975b, p. 17). Reflection discloses that the problem of generating dimension n is not a quantitative one of increasing the number of well-formed boundary elements in the manifold of D_n to infinity; it is a problem of creating the *quality* of boundedness in the first place. If a boundary exists that permits one sharply to distinguish any point from another, in effect, we already possess the full-blown, infinitely dense line; all that is required for the line is this quality of two-pointedness. Distinctions among numbers of points beyond two therefore are irrelevant to the question of dimensional generation. Thus I suggest that, rather than being composed of a finite number of well-defined points, a fractional line is a more "primordial" entity, one whose boundary element is not yet completely formed, has not yet fully been sharpened into focus. (This, of course, is the problem with which Menger [1940] was grappling when he abandoned his "topology of lumps" to contemplate "something still more vague.")

In sum, while Muses introduced the notion of "dimensional generation," he tacitly was accepting the classical intuition of invariant, integer dimensionality. In employing dimensions with already-formed identity elements, Muses failed to meet the primary requirement for an adequate concept of dimensional generation: one must account for the generation of the identity elements themselves.

The proposition I offer is that this requirement may be met by an approach that takes its cue from the properties of the Moebius surface. I suggest that when the Moebius operation is interpreted neo-intuitively (without imposing upon it the assumption of topological invariance) and properly generalized, it can provide a fruitful insight into the qualitative domain of transformation between dimensions.

3.3 The Primordial Phase of Dimensional Generation and the Role of Perspective

In the preliminary model of symmetry transformation given in section 3.1, I examined the effect of longitudinal Moebius expansion on two points (presumed initially well defined) in opposite sides of the surface, points bounding an interior line perpendicular to the idealized two-dimensional surface (see figure 5.2b). By taking line PP' as already generated, the expansion symbolized the *de*generation of this line, its loss of density, brought out in the progressive identification of P and P'. In this manner, a provisional account of spatial symmetry-breaking was provided, one that served the limited purpose of giving first expression to Menger's "topology without points." What we found in the previous section is that mounting a thoroughgoing challenge to the classical intuition of space requires us to think in terms of dimensional *generation*, of symmetry-*making*. This conclusion will gain further support when, in the course of articulating the Moebius concept, we see that it is indeed appropriate to view line PP' as not yet formed at the outset, as possessing "zero density"[4] *to begin with.* To launch our investigation, let us reconsider the relationship between P and P' at transverse section AB, before Moebius expansion has commenced.

From both classical and neo-intuitive perspectives, it is *possible* to say that in the pre-expansion circumstance, P and P' "do not exist as separate entities." However, from the classical standpoint, this statement is either trivialized or turned into a mere negation, essentially because continuity is taken as an either/or proposition. The classical prerequisite for continuity is clear: A continuum is an entity of integer dimension n that possesses a well-defined (fully formed) boundary element (of D_{n-1}). The statement that P and P' do not exist as separate entities is trivialized when initial state AB is viewed from "above," that is, in the three-dimensional frame of reference within which the boundary element of line PP' is already formed. Here, while we might imagine a "nondense" state of PP' in which P and P' are regarded as the "same point," the condition for the *infinite* density of the line (thus the sharp differentiability of P and P') actually is satisfied; no dimensional generation is required. (This defines the case in which the identification of P and P' is merely topological.)

If the classical approach is maintained, the only alternative to viewing pre-expansion state *AB* from "above" is to view it from "below." Now the assertion that *P* and *P'* do not exist as separate entities reduces to the idea that they do not exist at all, since, from the perspective of strict, *two*-dimensional continuity, the boundary element of the perpendicular interior line is simply nonexistent. Such a categorical separation of continua is in keeping with the Kantian intuition of dimensionality that makes unthinkable the concept of a domain *between* dimensions where space itself could undergo development. While dimensional generation is unnecessary when viewed from "above," when viewing it from "below," it is simply not possible. Were it necessary to make the transition from a continuum of D_n to one of D_{n+1} in the classical fashion, the latter would have to be created *ex nihilo*, to spring full-blown from the "brow of Zeus."

On the other hand, the neo-intuitive approach supersedes the binary constraint that if *P* and *P'* exist at all, they can exist only as fully differentiated elements of an already formed continuum. Intervening stages of development are implied. In the neo-intuitive formulation of their pre-expansion relationship, *P* and *P'* truly do not exist in separation, yet they do exist. This is the undifferentiated state of affairs adumbrated above, the condition defining the "primordial line." Here, though the perpendicular linear continuum (denoting extension in three-space) is not to be regarded as merely nonexistent, it is as yet unformed because the identity element that defines it *as* a continuum, the point boundary, has not yet been formed. To begin to grasp the meaning of this embryonic stage in the point's development as boundary of the line, we must hold in abeyance the classical intuition of the point that compels us to picture it as a sharply articulated entity. Neo-intuitively, the "primordial line," the line of "zero density," is not a line with a single, well-formed point but one whose point-element is *formless*. I propose that this is the circumstance that defines the initial, pre-expansion phase of dimensional generation and that through expansion of the Moebius kind, the continuum is generated via differentiation of its boundary.

Of course, before this can be shown, the neo-intuitive interpretation of the pre-expansion state must be better understood; an adequate basis for it must be provided. I begin by offering phenomenological support for a critical distinction not recognized in the traditional analysis of dimensionality, that between a continuum already generated and one in the process of being generated. Presently to be demonstrated is the uniqueness of the *third dimension* as it relates to our experience.

Classically, continua of all dimensions must be treated in the same way, presupposed to be fully generated. Yet, on ordinary phenomenological grounds, a clear-cut distinction can be drawn between three-dimensional structures and entities of lower dimensionality. The points delimiting a segment of a line and the lines bounding a section of two-space are simultaneously perceptible; they can be expressed in a spacelike fashion, made copresent for measurement. On the other hand, the surfaces bounding a volume of three-space must be viewed in *perspective*. At a given moment, some of the surfaces necessarily will be concealed from immediate perception, making the total surface area observable only by a sequential operation. A timelike interval will be required for complete observation of the surface of a solid.

The classical assessment of this phenomenological distinction may best be understood by restating it in terms of the "internal structure" of the *single* boundary element. One could observe, for example, that while the opposing edges of a line bounding a plane are simultaneously perceptible, opposite sides of a surface are not (a fact illustrated by the impossibility of viewing simultaneously points P and P', situated in opposite sides of the surfaces depicted in fig. 5.2). However, when the line is viewed strictly in its capacity as boundary element of two-space, it actually is inaccurate to speak of its "opposing edges." From the classical standpoint, we would be no more entitled to speak of "opposite sides" of the plane, taken as the boundary element of three-space. This is because the intuition of continuity implies the *unextendedness* of all boundary elements. The element, being "infinitely thin," devoid of intrinsic structure, cannot have opposing edges or sides. Classically, the boundary element is to be regarded not as internally oppositional but as *juxta*positional, distinguished only by its external relatedness to its neighboring elements in the densely packed continuum.[5]

The classical presupposition of the juxtapositional character of the plane is none other than Poincaré's: the plane, taken as a boundary element, must be well defined as such, so that it bounds the three-dimensional continuum just as sharply as the line bounds the plane and the point bounds the line. (With respect to line PP' of fig. 5.2, if we compare this extension in three-dimensional space [containing points in opposite sides of a two-dimensional surface] with a line that does not implicate the third dimension [such as line AB, inscribed in a single side of the surface], on classical grounds, no intrinsic distinction could be made. The point-elements of both lines would have to be regarded as simply juxtaposed, the continua as infinitely dense.)

Of course, though oppositional properties such as sidedness are considered to have no effect on the boundedness of space, no bearing on its local character, they are not simply disregarded in the classical approach. Instead they are interpreted as *topological* features (as noted in section 3.1). The sidedness of an entity refers to the global form taken by the entity when it is embedded in a locally juxtapositional space. This notion of embeddedness accords with the aim of maintaining the fixed disjunction between continua. To say that the boundary element of a continuum of D_n is juxtapositional is to say that the dimensionality of this element is precisely $n - 1$ (being utterly unextended in the space of D_n). What is the dimensionality of the highest dimensional structure *embedded* in n-dimensional space? Exactly one dimension less. That is, it will possess D_{n-1}, but it will do so only globally, the dimensionality of its own juxtapositional boundary element being $n - 2$.

Although the phenomenological uniqueness of the planar boundary element loses its significance on the classical view, in the context of dimensional generation it plays a central role. The neo-intuitive approach grants that the *line* may be viewed as a fully formed boundary element of two-space, in which case its opposing edges reduce to internally structureless lines in mere juxtaposition. Yet the sides of the plane are neo-intuitively interpreted as entailing true internal opposition, an intrinsic structuring that confounds the classical dichotomy between the local and topological properties of space.

Classical intuition precludes the idea of intrinsic structure. The term *intrinsic* (defined in the sense given by endnote 2) applies only to the structureless relation prevailing locally, whereas *structure* is a term reserved for attributes that are exclusively *ex*trinsic, that is, topological. The merely extrinsic character of the linear structure is borne out by the observation that its opposing edges, being simultaneously perceptible, *are* readily reducible to structureless lines in juxtaposition. In contrast, the simultaneous *non*perceptibility of the sides of the plane, their phenomenological nonjuxtaposibility, calls the extrinsicality of the planar structure into question. This is the phenomenological fact that is bypassed in the classical approach. The notion that the oppositional structure of the plane is but a topological feature of a globally two-dimensional entity embedded in locally juxtapositional three-space is maintained by an act of abstraction in which nonjuxtaposibility is denied. The neo-intuitive approach, in *adhering* to phenomenological fact, arrives at the contrary conclusion that the structure of the plane indeed is dimensionally intrinsic.

Now, whether the classically idealized plane is viewed as a fully formed boundary element of three-space or a merely global entity embedded in three-space, its dimensionality is two. On the other hand, the actual, perspectival plane can be said neither to bound sharply nor to "lie in" higher-dimensional space while retaining its own strictly lower-dimensional character. The neo-intuitive concept to be grasped is that three-dimensional space is *incomplete, is being generated*; the perspectival quality of the plane is indicative of an intermediary stage in this generative process. Thus, viewed in global terms, the plane is not of D_2 but is of $D > 2$, is "more than" two-dimensional; it is not just embedded in already formed three-space but is three-space itself in *partially* generated, "fractional" form. Taken as a local entity, the dimensionality of the perspectival plane is "less than"[6] two; it is not yet a structureless boundary element of three-space, not yet juxtapositional (as phenomenological observation suggests), since the continuum is still in the process of being differentiated.

What permits us to treat three-space classically, as if it were already fully formed? The implicit "spatialization of time."[7] That is, the timelike interval required to measure the surface area enclosing a volume is interpreted as if it were independent from a nontemporal, fully generated three-space, rather than the factor indispensable to making three-dimensional measurement possible in a space that is *not yet* three-dimensional. In this unacknowledged appropriation of time for space, the inherent temporality of three-dimensional space is overlooked.[8]

Now, while the perspectival state of affairs is neo-intuitively understood as indicating the incompleteness of three-space, it does not represent the *most* primordial phase of dimensional development. In the latter, the plane as boundary element of three-space would be *entirely* undifferentiated, affording no sense whatever of opposing sides. Here I am speaking of the "zero density" condition alluded to above, the condition proposed to exist prior to the Moebius-type expansion. Having examined the question of perspective, the problem of truncation mentioned earlier (p. 78), in connection with the provisional model of Moebius expansion, now can be understood. The account given in section 3.1 is truncated because, in it, the primordial phase of expansion is approached from the less primordial standpoint of perspective.

In the preliminary illustration, the Moebius transition to one-sidedness was seen to involve a progressive merging of points that initially were maximally differentiated, points that occupied positions in diametrically opposing sides of the surface. The neo-intuitive insight presently at hand is that the initial phase should not be viewed as a differentiated, two-sided one but as an utterly *un*differentiated state in which the sense of opposition between points P and P' has not yet developed. Although the viewer, already possessing perspective, will observe P and P' in a common frame of reference in which their opposite sidedness is readily discernible (and, if classically inclined, he will assume them merely to be juxtaposed), the *pre*perspectival points themselves may be thought of as "infinitely distant" from each other.

The term *infinite* is to be understood here in the sense of a *closed* infinity, not a merely open one. The latter would be associated with increasing the metrical distance between well-defined points to an arbitrarily large (yet essentially finite) value, an operation in which the distinct identity of the points would never be in doubt. In the former case, P and P' would share the *same* identity: the "endpoints" of a closed infinite line are identified. In the neo-intuitive interpretation of this mathematical concept, the sameness of points is not trivialized by misconstruing the line at infinity to be well-differentiated, as a finite line segment would be. Rather, the line at infinity is the undifferentiated, primordial line.

The neo-intuitive meaning of the separation and unity that would prevail at infinity bears some discussion. P and P' would be regarded as united, not in the sense of distinct entities existing together, but in their nondistinctness as such. Here the difficulty for classical intuition again becomes evident, for we must apprehend the notion that at infinity, P and P' are not simply actualized (well-differentiated) identities but "more than" mere potentialities (not-yet-existing possibilities). However, the idea of a domain between actuality and potentiality, of "actual potentiality," is not without precedent. According to Stapp (1986, p. 265), an intuition of this sort was first suggested by David Bohm (1951) in his characterization of the microphysical wave function as indicative of "tendencies" or "potentia." Following Heisenberg, who endorsed Bohm's insight, we might say that primordially, P and P' exist in "a strange kind of physical reality just in the middle between possibility and reality" (1958, p. 41). It is with respect to the "actual" aspect of the "actual potentiality" that P and P' would be united: they would exist in nondistinctness. Their separation would be associated with the "potentiality" aspect: they would not *yet* exist as distinct. The latter is the nonmetrical, qualitative sense in which P and P' would be "infinitely distant." *[See the discussion of infinity in chapter 4.]*

Before we can proceed to examine the effects of Moebius expansion on the primordial relationship between P and P', we must come to grips with a major issue deferred until now: the need for a higher-order expression of the Moebius process. We are about to see why expansion in the Moebius strip as such, in fact cannot account for the generation of *three*-dimensional space. The actual role of the Moebius surface in the concept of dimensional generation will be clarified later on.

3.4 Generalization of Moebius Process

The classical assessment of both the Moebius strip and its cylindrical counterpart is that they are two-dimensional entities topologically embedded in juxtapositional three-space. When the concept of dimension is made answerable to the phenomeno-logical facts of perspective, it becomes evident that neither entity rightly can be thus regarded. Nevertheless, we have found that it is possible (by a "spatialization of time") to *assume* three-space to be a fully developed continuum, and this enables us to treat both of the entities in question in the classical manner.

The possibility of such an idealization of the Moebius strip means that expansion in it has no intrinsic consequences for three-space. Stated in customary terms, the Moebius strip is a "properly constructible form": Like the cylindrical ring, it can be put together in three-dimensional space by a sequential operation that does not tear it, that is, by a continuous transformation, one permitting the assumption that topological invariance is maintained. In constructing the surface, no effect on spatial symmetry is in evidence.

However, a more general version of the Moebius surface exists that is not "properly constructible": Klein's surface or the *Klein bottle* (fig. 5.4b).

Figure 5.4

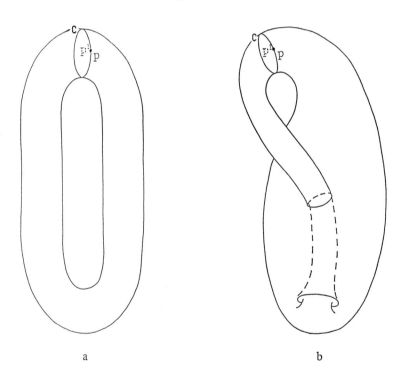

a   b

Torus (a) and Klein bottle (b)

Figure 5.4 is a higher-order analogue of figure 5.2. The relationship shown between the Klein bottle and the torus (fig. 5.4a) reflects that between the Moebius surface (fig. 5.2b) and the cylindrical ring (fig. 5.2a). The two-sided torus is a closed surface formed by joining the end circles of an elongated open cylinder. In forming the one-sided Klein bottle, end circles are not brought together by a continuous operation in the three-dimensional space exterior to the surface; they are superposed by entering the region *interior* to it, an operation requiring the surface to pass *through* itself. This self-intersection (which is the counterpart of the twist necessary to produce the Moebius strip) entails an unavoidable breach in the surface, a loss of continuity attesting to the fact that in the case of the Klein bottle, the symmetry of three-space cannot simply be assumed to be maintained. Here an intrinsic transformation of spatial symmetry is indicated. To be sure, such an interpretation would not accord with the conventional evaluation of the Klein bottle. In the latter, classical appearances are saved by resorting to higher-order abstraction, as we shall see. For the present, let us pursue the conclusion we have been led to: It is expansion in the Klein bottle, not the Moebius strip, that can express the generation of three-space.

In figure 5.4b, longitudinal expansion is imagined to begin at the cross-section of the Klein bottle bounded by circle C. Viewed perspectively, this defines the two-sided limit case in which opposite sides of the surface are simply differentiated. The sense of opposition once more is given by points P and P', one situated inside the bottle in the circumference of C, the other in the corresponding position on the outside; again it may be assumed that P and P' bound an interior line associated with the thickness of the surface, its extension in the dimension perpendicular to itself (to keep the figure simple, line PP' is not shown).

Of course, when this perspectival way of viewing the pre-expansion state is grasped neo-intuitively, it is itself seen to represent an *intermediary* phase of dimensional development; the initial phase actually entails the *pre*perspectival condition of "infinite separation" in which the experience of inside and outside, of opposition between P and P', has not yet begun. This primordial Klein bottle corresponds to the incipient stage in the generation of three-space. (The initial primordiality of the Klein bottle will receive direct confirmation, when, in the concluding portion of section 3, it is seen to arise as an immediate consequence of the developmental relationship between the Klein bottle and the Moebius surface.)

Commencement of Klein-bottle expansion to one-sidedness initiates the merger of *P* and *P'*, "reducing" their "infinite separation," creating the common frame of reference necessary for experiencing their mutual opposition. *P* and *P'* emerge from primordial unity to become distinct entities, perspective arising. Thus, while longitudinal expansion in the two-sided torus would leave *P* and *P'* in their "infinitely distant," undifferentiated state, Klein-bottle expansion toward one-sidedness effects their differentiation. Note the implication of this: when experienced perspectivally, both the torus and the Klein bottle are differentiated structures, but only the latter can be said to differentiate *itself.* Such an operation is associated with an increase in the density of line *PP'* from its primordial value of zero to a finite term. In the perspectival state attained, dimensionality becomes "fractional," globally "greater than" two but "less than" three, locally "less than" two but "greater than" one. Then, carrying Klein-bottle expansion to completion, the density of line *PP'* would become infinite, the dimension number of the space in which it would extend assuming integer value (globally three, locally two). In this terminal stage of development, perspective would be overcome, Klein-bottle one-sidedness making both sides of the plane simultaneously perceptible for the first time. Thus the criterion for classical continuity apparently would be met, the generation of spatial symmetry completed. Yet the climactic phase of dimensional generation would be decidedly *non*classical in character.

Though the Klein-bottle transformation does not entail a simple breaking of spatial symmetry, neither does it involve simple symmetry-making. Above, it was established that in a classically symmetric continuum, intrinsic structure is prohibited. This means that an identity element of space can possess no boundary that would separate an interior region of it from what would lie outside; *all* must be "on the outside," as it were. In other words, locally, classical space consists not of internally substantial, concretely bounded entities but only of abstract boundedness as such. Sheer externality alone prevails, the " 'outside-of-one-another' of the multiplicity of points" (Heidegger 1962, p. 481). This is the meaning of juxtaposition. In contrast to this, the primordial condition would be one of utter bound*less*ness, sheer internality, the purely concrete "inside-of-itself of the unbounded point" (to transpose the phraseology of Heidegger).

Now, the Klein-bottle expansion by which the condition of boundedness is created is not a process of mere differentiation. It is not sheer juxtapositional diversity that is produced by this movement "away from" primordial unity but a diversity-in-*unity*. For differentiation of *P* and *P'* occurs in the course of the transition to one-sidedness that brings them back *together*, now *as* distinct entities. While the distinctness of *P* and *P'* needs to be established (in the intermediary phase of expansion) before unification can fully be realized, said differentiation does not occur as a separate process but is inextricably linked to the unification process itself. Thus, the Klein-bottle transformation is not differentiative or unitive alone, or a linear combination of these, but is both differentiative and unitive at the same time.

The climactic phase of the Klein-bottle operation should be especially problematic for classical intuition: we need to intuit a state of complete boundedness (differentiative aspect) that at once is entirely unbounded (unitive aspect). In the intermediary phase of the transformation, when the distinctness of boundary elements still is in the process of becoming established, *P* and *P'* are neither fully differentiated nor wholly identified. In the terminal phase, they would be both! Though elements now would be simultaneously perceptible, their concrete opposition would not simply have collapsed. *P* and *P'* would not merely be juxtaposed, abstractly outside of each other. They would be completely outside *and* inside at the same time. (As we shall see later, the space generated in this way indeed will take on all the characteristics of the classically juxtapositional continuum, when the developmental process is carried still further.)

3.5 The Principle of Synsymmetry

Faced with the neo-intuitive challenge of comprehending a stage in the development of space when its identity elements are both utterly distinct and, at the same time, wholly identified, we may gain assistance from the field of perceptual phenomenology.

In figure 5.5, the three phases of dimensional generation are depicted. The preperspectival phase (fig. 5.5a) is that in which the symmetry of three-space is completely undeveloped.[9] In the second phase (fig. 5.5b), distinct perspectives arise, suggesting the emergence of "depth perception." This symbolizes the fact that space has attained finite density, that it has been partially

Figure 5.5

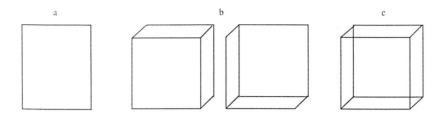

Phases of dimensional generation: preperspectival (a),
perspectival (b), and postperspectival (c)

generated. The final phase of dimensional generation is portrayed
by figure 5.5c, a figure known in phenomenology as the Necker
cube *[see chapters 1 and 3]*.

The Necker cube merges the perspectives of figure 5.5b,
incorporates them in a single spatial configuration. Yet, in our
customary way of viewing the cube, we perceive only one of these
perspectives at a time; thus, with respect to *time*, the cube
remains polarized. Nevertheless, it is possible to view the cube
in a thoroughly integrated fashion. Rather than allowing our
experience of the figure to oscillate from one perspective to the
other, we may apprehend both perspectives at once. If this is done
without allowing the cube simply to flatten into an array of
connected lines, that is, if the awareness of depth is retained, one
will experience a sense of self-penetration: the form will appear
to go *through* itself. Such a mode of imaging has a revealing effect
on the perception of the cube's faces.

In the conventional, perspectivally polarized way of viewing
the figure, when the transition is made from one pole to the other,
the three faces of the cube that were seen to lie "inside" presently
appear on the "outside," and vice versa. But it is only at "polar
extremes" that faces are perceived as *either* inside *or* outside.
With perspectival fusion, each face presents itself as being inside
and outside at the same time, thereby producing an experience
of *one*-sidedness, one in which all six faces—all the bounding
surfaces that enclose the three-dimensional object being
symbolized—are simultaneously given. It should be emphasized
that this self-intersecting experiential structure does not merely

negate the perspectival distinction between sides, reproducing preperspectival flatness. In the postperspectival state, faces *are* inside, yet they are *outside* as well. Thus the structure is as fully two-sided as it is one-sided.

The two forms of figure 5.5b, representing sets of diametrically opposed surfaces bounding a volume of three-dimensional space, are entirely analogous to points *P* and *P'* of figures 5.2 and 5.4. And the integration of these forms provides the needed insight into the climax of Klein-bottle expansion, wherein boundary elements are both utterly distinct (since each is now fully differentiated, being perceptible in its entirety, completely unconcealed) and wholly identified. Furthermore, while classical intuition would have us picture *P* and *P'* as points in the sense of being simply self-symmetrical and extensionless, thus entities without intrinsic structure, the present formulation brings to light the underlying *chirality* (handedness) of the identity elements of the space being generated, the specific nature of their *dis*symmetry. In the course of dimensional generation, the boundary elements of the space whose symmetry is being made are related to one another *enantiomorphically*, as asymmetric mirror twins.[10]

To understand better the role of symmetry in the process of dimensional generation, let us return once more to the traditional definition: A symmetry is identified when a characteristic of a body or system remains the same despite the fact that a change has been introduced. To repeat my earlier example, if a sphere is transformed by rotating it through some angle about its center, its appearance will not change. The sphere is therefore symmetric under the operation of rotation. *Spatial* symmetry, as I have defined it, implies that the sphere or any other geometric object would remain invariant under any transformation that does not disrupt the juxtapositional, locally continuous character of its identity elements, that is, that does not require a loss of extensive continuity. From the abstract perspective of the continuum, all concrete variations in the appearance of the object would "look the same." Applied to the members of the enantiomorphic pair shown in figure 5.5b, because one perspective readily can be transformed into the other by a continuous rotation in three-space (as can be demonstrated by rotating any perspectivally viewed cubelike object through an angle of 180°), the enantiomorphs would be regarded as perfectly symmetric. This is equivalent to interpreting the opposition of enantiomorphs as a merely topological feature of structures *embedded* in locally continuous space: concrete oppositional asymmetry is reduced to abstract juxtapositional symmetry.

Of course, a distinctly different interpretation of enantiomorphic asymmetry is offered in the *non*topological context of dimensional generation. We have seen that, neo-intuitively, the perspectival plane is to be regarded not as a merely topological entity embedded in an already formed three-dimensional space but as the entity involved in the intermediary stage of that continuum's development. Now, in coming to appreciate the particular asymmetric character of perspective, we recognize the intermediary stage as one in which merging enantiomorphs (P and P') have emerged from an undifferentiated, pre-asymmetric condition and entered into intrinsic opposition. In the final stage, the postperspectival culmination of enantiomorphic fusion, concrete opposition is overcome, not simply by being dissolved or bypassed, but by being brought to full expression. This is the paradoxical Klein-bottle state apprehended through the perspectival integration of the Necker cube: a totally differentiated two-sidedness at once is intuited as an utterly unitive one-sidedness. With this insight, the dialectical nature of dimensional process is clearly disclosed. The dichotomy between pre-asymmetric "thesis" (unitive aspect) and asymmetric "antithesis" (aspect of diversity) is surmounted in an achievement of *synthesis* (a unity of unity and diversity). Accordingly, the climactic mode is characterized as *syn*symmetric *[see chapter 2]*. The fact of dimensional generation to be clarified in due course is that the radically nonclassical Klein-bottle transformation of P and P'—from pre-asymmetric to asymmetric to synsymmetric relatedness—is followed by their entry into classical juxtaposition. We will see that spatial symmetry arises from the synsymmetric union of pre-asymmetric and asymmetric phases of dimensional development.

3.6 Time and the Generalized Moebius Sequence of Dimensional Development

The enantiomorphs of figure 5.5 can be given proper classical treatment only as long as they are viewed in the polarized fashion, as categorically distinct perspectives. Postperspectival *fusion* of enantiomorphs should make classical analysis problematic, since, in the Klein-bottle self-intersection it entails, the discontinuity of three-space would be undeniable. Of course, the strikingly nonclassical character of the Klein bottle is no revelation to mathematical physics. How is the Klein bottle conventionally approached? By resorting to a higher order of abstraction, so as to save classical appearances. That is, the consequences of self-intersection are circumvented, the traditional presupposition of continuity retained, simply by assuming the Klein bottle to be embedded topologically in an "extra dimension of space." Whether the significance of this "hyperdimensional continuum" is presumed to lie only in the mathematical convenience it provides, or it is interpreted to have "real existence" (as claimed for the "primordial manifold" of current Kaluza-Klein theory), it remains an extrapolation of classical thought. On this view, one might speak of a Klein-bottle fusion of enantiomorphs by expansion through "perpendicular four-space," and this transformation would be topologically continuous in its abstract way, since the self-intersection would be eliminated.

Now, the neo-intuitive approach also accepts the idea that an additional dimension is implicated in the Klein-bottle transformation. But the extra dimension is not simply invoked in a stratagem for upholding classical continuity. In the concept of dimensional generation (which is concerned with the transformation *of* continuity, thus not limited to continuous transformations), the dimension inherently accompanying the generation of three-space would be construed as undergoing its own qualitatively distinct development. I am referring to the transformation of the dimension we experience as *time*.

As noted twice before, the limitative nature of perspectival experience is classically offset by the "spatialization of time." That is, when measurements in three-space are made, time must be used to compensate for the actual incompleteness of space, but instead of acknowledging this, space is taken as simply complete and time is treated independently. Thus, while time is appropriated to render space a finished product, it is presumed to retain its own entirely distinct character, a quality of pure succession independent from and antithetical to the juxtapositional simultaneity that is space. In the classical "spatialization of time," the categorical *separation* of space and time is maintained, in keeping with Kantian intuition. (Note that the mathematical unification achieved in the relativity theory leaves the *qualities* of simultaneity and succession foreign to one another.)

From the neo-intuitive standpoint, space and time bear a more intimate relationship. Rather than being viewed as pure forms, they are understood as *mutually complementing hybrids*. The incompleteness of spatial experience found in the perspectival stage of development (the impossibility of simultaneously perceiving opposite sides of surfaces) is interpreted as the "timelike" aspect of the spatial hybrid. Similarly, time would not involve sheer succession but would possess a "spacelike" feature. With respect to the latter, it can be shown that if time did conform to its classical idealization, it actually could not be appropriated to treat space as though it were complete. Consider as an illustration the sequential operation necessary for observing points P and P', situated in opposite sides of a surface. Point P is recorded at time T, point P' at T'. If the relation between T and T' were one of *sheer* succession, if there were no temporal overlap sustaining a trace of the first event of measurement for the occasion of the second, P and P' would be incommensurable. This complementing feature of overlap (temporal simultaneity) that fills the gap created by spatial nonsimultaneity is the spacelike component of the temporal hybrid. (Of course, in the Cartesian idealization of space and time, both spatial nonsimultaneity and temporal simultaneity are considered "merely psychological effects" divorced from physical reality itself.)

Let us now attempt to comprehend the relationship between space and time over the full course of its development. I propose that the transformation of time follows the same basic dialectical pattern as described for space, but with a certain reversal in the order of events.

Consider again the primordial state of affairs in which spatial identity elements P and P' exist in "actual potentiality": together in nondistinctness (actuality aspect), "infinitely separated," disjoined *as* well-defined units of three-space (potentiality component). With ensuing Klein-bottle expansion, potentiality is developed and P and P' begin to be differentiated in oppositional perspective, the nonsimultaneity of which constitutes the timelike feature of the spatial hybrid. By the climactic phase, the two-sided opposition of P and P' has increased to a maximum, yet in the same process, a state of *one*-sidedness also has been attained in which P and P' have become simultaneously perceptible (they are both nonsimultaneous and simultaneous at once!). The latter aspect of the synthesis might be said to entail a "lifting of the repression" on the unitive actuality state wherein P and P' were primordially identified. The simultaneous perceptibility of P and P' *as distinct elements* satisfies the condition for the subsequent emergence of three-space as a classical, nonoppositional continuum, one in which the timelike component has vanished.

What may we say of the complementary development of the fourth dimension? Evidently, the primordial state of T and T' would involve "infinite *temporal* separation." To decipher this, we must bear in mind that separation is a quality that is *inherent* to time, since its essence is succession (whereas space is associated with nonseparative simultaneity). This suggests that the components of the "actual potentiality" have a reversed meaning for time: infinite separation is *actual*, nonseparation is potential. If this is correct, the transformation of time would run its course in the direction opposite to that of space.

For time, the potentiality that would begin to be realized with the onset of expansion would be the *de*differentiation of T and T'; this is the temporal "overlap" referred to above, that which forms the spacelike feature of the temporal hybrid. By the climactic phase, the nonseparation of T and T' would have reached its maximum, overlap would have become total; yet it also would be the case that in this stage "repression" would have been "lifted" on the primordial actuality of infinite separation. What condition would be satisfied here? That of an incipient, fourth dimension of space! I propose that the utterly dedifferentiated T and T' constitute themselves as nascent *spatial* opposites existing together in nondistinctness, and that the resurgence of infinite separation is their potential for differentiation in a subsequent epoch of development. (Note how actuality and potentiality would exchange roles in the transformation of time into space.)

The outcome envisioned is wholly consistent with the interpretation of dimensional development as a process of enantiomorphic fusion, for it is clear that the very same process that resolves the opposition of figure 5.5b spatial enantiomorphs *creates* the condition for higher-order spatial opposition. This is shown in figure 5.6, where it is demonstrated that the Necker cube produced from the merger of lower-order enantiomorphs is itself a member of a higher-order enantiomorphic pair. Summarized

Figure 5.6

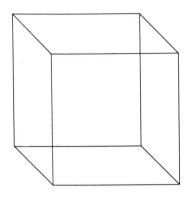 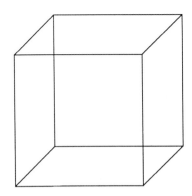

Necker-cube enantiomorphs

in general terms: by the same dialectic in which a space of given dimensionality is brought to maturation, a higher dimension of space is incubated through the transformation of time.

Time may be incorporated more fully into the theory of dimensional generation by expressing it in the idiom of dimensional valuation established earlier. By virtue of the fact that it possesses a spacelike component, it may be said to have a certain "density"; yet unlike the positive density of three-dimensional space (and spaces of lower dimensionality), the density of time may be regarded as "negative." This accords with Muses's (1985) view that time, when qualitatively understood, is associable with "negative dimensions—which are not *extension* as are [positive] spatial dimensions, but *intension*" (p. 8). In this conceptualization of dimensional density, the Klein-bottle expansion, which raises the positive density of three-space from a primordial value of zero to an infinite value, also would involve an increase in the density of the fourth dimension from an infinite negative value to a finite negative value (the familiar, perspectival experience of time) to a value of zero. A true spatialization of time would be the result. And while the classical "spatialization" serves the interest of rendering dimensions changeless, the nonclassical alternative would imply a *continuation* of generative process. Coincident with completing the development of three-space, a new, as yet undeveloped, fourth dimension of space would be brought to term, to emerge in perspective and undergo its own spatial maturation in a subsequent epoch of dimensional generation. In light of the generalized account of dimensional transformation that now has been provided, a fuller understanding of the role of the Moebius surface becomes possible.

In section 3.3, I took up the question of perspective to demonstrate the phenomenological uniqueness of three-dimensional space relative to lower dimensionalities. We experience the latter nonperspectively, as fully generated, infinitely dense continua. But this does not mean that they should be viewed as never having been otherwise. The theory of dimensional generation requires that spaces of $D < 3$ also were subject to dialectical process, that they have undergone their own generation in *previous* epochs of development.

This realization leads to a deeper grasp of why the Moebius transformation does not affect the symmetry of three-space. It is not the current epoch of development in which synsymmetric consequences of Moebius expansion are felt but the *previous one*, the epoch in which *two*-dimensional space was being generated and time was experienced as the *third* dimension.

In the present epoch, where the boundary element of the space being generated is the surface, the condition of perspective that must be surmounted synsymmetrically is that in which opposing sides of the surface cannot be perceived simultaneously. Expansion to Klein-bottle one-sidedness can achieve this transformation whereas Moebius-strip expansion cannot; the one-sidedness of the Moebius is nonintrinsic, merely perspectival. On the other hand, in the previous epoch implied by the theory, when two-space was being generated, the condition of perspective to be transcended was that in which opposing *edges of the line* were not simultaneously perceptible. Evidently, this was the epoch in which the influence of Moebius expansion was dimensionally intrinsic.

In addition to being one-sided, the Moebius surface is one-edged. That is, it not only identifies opposite sides of itself (nonintrinsically) but merges opposing edges, as well (in contrast to expansion in the cylindrical ring, in which the simple distinctness of edges is maintained). This Moebius fusion of edges, carried out in the *prior* epoch of dimensional generation, must have played precisely the same role as is presently played by the Klein-bottle fusion of sides. (Note that in that Moebius epoch, one could not have spoken meaningfully of opposite sides of a surface, any more than one now can speak of the "opposite sides" of space.) In the two-dimensional space then existing, the Moebius strip clearly would have been a self-intersecting form, thus not a "properly constructible" one. No such structure could have been put together by a continuous transformation that would have preserved the assumption of topological invariance, of spatial symmetry. Of course, on the neo-intuitive account, the Moebius transformation is associated with *generating* the symmetry of two-space, in the same synsymmetric manner the Klein-bottle operation is seen to act on three-space.

To underscore the thoroughly fluid character of the developments being described, it may be noted that no singular, process-interrupting point of transition is required in passing from one epoch to the next. As I indicated above, the enantiomorphic merger performed in epoch n is at once an incubation of epoch $n + 1$ enantiomorphy. Thus it can be said that, in the previous epoch, the Moebius expansion that fused the "linear enantiomorphs" of two-space at the same time was creating the Moebius surface itself as the enantiomorphic structure to be fused with *its* opposite in the Klein-bottle generation of three-space (the enantiomorphs of fig. 5.5b are taken to represent oppositely oriented Moebius surfaces whose synsymmetric merger yields the Klein bottle). This intimate synchronization of processes implies that the climactic, postperspectival phase of the last epoch of expansion must have constituted the primordial, preperspectival phase of the present epoch, with "nothing in between." Here D_2 would have reached infinite positive density, the edges of its linear boundary element becoming simultaneously perceptible for the first time; D_3 would have gone from finite negative density to a density of zero, creating the nascent experience of three-dimensional space from what had been experienced as time; and D_4 would have emerged as an entity of infinite negative density, indicating the rudimental appearance of a new, fourth dimension of time. In this primordial state, the newly conceived Moebius enantiomorphs would have been utterly undifferentiated, "infinitely distant" as potentially distinct entities. And accompanying these embryonic spatial enantiomorphs would have been utterly *differentiated* temporal enantiomorphs, oppositely oriented *Klein bottles* (analogous to the forms of fig. 5.6) in their "hyperprimordial" state, their "infinite separation" being *actual.* Then the present epoch would have begun to unfold in the 3 + 1-dimensional perspective we now experience. In this regard, it should be mentioned that only now is the two-space that underwent synsymmetric transformation in the previous epoch, simply symmetric. Thus we may surmise that a space will take on classical continuity only in the intermediary phase of the epoch following that of its generation.

3.7 Quasi-Mathematical Summary of the Generalized Moebius Concept of Dimensional Development

The essentials of dimensional generation have been set forth. To be sure, additional work is required in order to bring the idea to fuller expression. But work of what sort? Is further development to be defined in terms of a rigorous quantification of the qualitative notions that have been offered? The problem is that quantification itself is an activity that gives priority to mathematical invariance, to *symmetry.* Serving the aim of classical intuition, the enterprise of quantification should not be viewed as constituting a neutral, universally applicable method. If we sought merely to quantify the notion of synsymmetry, in effect, we would be attempting to reduce it to symmetric terms, and this would contradict its essential meaning. The synsymmetry notion *supersedes* the classical procedure of reducing concrete particulars to abstract generality. For synsymmetry is a concrete *generality* within which abstraction itself (mathematical symmetry, quantification) arises as but a limit case. Then mathematics as such surely cannot be the means by which the idea of synsymmetry may be further developed and secured. Implying as it does a new intuition of dialectical process, synsymmetry cannot be grounded in the narrower, classical intuition; it can be brought to clearer, fuller expression only by approaching it on its own dynamic ground, appealing to it from within itself for self-clarification (as exemplified by the phenomenological exercise with the Necker cube). Given the role of the mathematical *within* the intuition of synsymmetry, clarification of the latter must shed light on the former. But synsymmetry cannot be mathematized.

Nevertheless, in the interest of following through on my usage of quasi-mathematical terms such as "density" and "fractional dimensionality," and of presenting the developmental scheme of section 3 in a more concise form for the reader accustomed to mathematical rendition, the theory of dimensional generation is summarized in table 5.1. The caveat to be borne in mind is that the mathematical symbols employed should not be understood in a merely formalistic sense, as empty abstractions whose only purpose is to resolve ambiguity through quantification. They are to be regarded instead as *metaphors* that point to the mathematically *irreducible*, concrete process of metamorphosis that has been adumbrated.

I_X is the integer dimensional (fully generated) spatial structure of epoch X that was brought to maturation in the previous epoch, Φ. D_{I_X}, the dimensionality of I_X, is given by n; c_n denotes the density of D_{I_X}, which reached a maximum value of $+\infty$ in the terminal phase of Φ, thenceforth being maintained at that maximum.

F_X is the spatial structure of epoch X possessing "fractional dimensionality," being generated in the course of X. D_{F_X}, the dimensionality of F_X, is given by p, a value with a coefficient, ϵ_X, whose exponent is seen to increase from the primordial to the terminal phase of the epoch, along with changes in dimensional density (c_p). According to Muses (1977), ϵ is a "hypernumber" (a nonreal number defined as $\sqrt{+1}$), which, when multiplied by itself, constitutes the algebraic equivalent of a hyperdimensional reflection: a "successive rotation, through a four-dimensional space and then back into ours," which transforms chirality, changes "a right hand into a left hand, and vice versa" (Muses 1977, p. 77). Neo-intuitively interpreted, the squaring of ϵ is equivalent to the enantiomorphic fusion resulting from expansion in the Klein bottle (where $p = 3$). Therefore, raising coefficient ϵ_X to the second power can be viewed as generating F_X, an operation associated with changing its dimension number to a *real* integer value, since $\epsilon^2 = (\sqrt{+1})^2 = +1$. (While it could be said that "$D_{F_X} = f(\epsilon_X^X)$," we know that the "exponential curve" for this function would not be topologically continuous, since said "function" actually would entail a qualitative transformation *of* continuity.) When D_{F_X} has thus been "raised" from $\epsilon_X(p)$ to $|+1|(p)$, F_X becomes I_{Ψ_2}, an integer structure of epoch Ψ (I_{Ψ_1} is the integer structure of Ψ that was generated in Φ). (Use of the absolute value of ϵ^2 and, below, of $i^2 = |-1|$, expresses and is justified by the fact that no qualitative difference exists between the real numbers $+1$ and -1, whereas hypernumbers like ϵ and

Table 5.1

Quasi-mathematical summary of generalized Moebius concept of dimensional development

EPOCH Φ		EPOCH X		EPOCH Ψ
Terminal Phase	**Primordial Phase**	**Intermediary Phase**	**Terminal Phase**	**Primordial Phase**
$D_{F_\Phi}=n \equiv$ $\;$ $c_n=+\infty$	$D_{I_X}=n$ $\;$ $c_n=+\infty$	$D_{I_X}=n$ $\;$ $c_n=+\infty$	$D_{I_X}=n \equiv$ $\;$ $c_n=+\infty$	$D_{I\psi_1}=n$ $\;$ $c_n=+\infty$
$D_{T_\Phi}=|-1|(p(\varepsilon_X)) \equiv$ $\;$ $c_p=0$	$D_{F_X}=\varepsilon_X(p)$ $\;$ $c_p=0$	$\varepsilon^2_X(p)>D_{F_X}>\varepsilon_X(p)$ $\;$ $+\infty>c_p>0$	$D_{F_X}=\varepsilon^2_X(p)=|+1|(p)\equiv$ $\;$ $c_p=+\infty$	$D_{I\psi_2}=p$ $\;$ $c_p=+\infty$
	$D_{T_X}=\iota_X(q(\varepsilon_\psi))$ $\;$ $c_q=-\infty$	$\iota^2_X(q(\varepsilon_\psi))>$ $D_{T_X}>\iota_X(q(\varepsilon_\psi))$ $\;$ $0>c_q>-\infty$	$D_{T_X}=\iota^2_X(q(\varepsilon_\psi))=$ $|-1|(q(\varepsilon_\psi))\equiv$ $\;$ $c_q=0$	$D_{F\psi}=\varepsilon_\psi(q)$ $\;$ $c_q=0$
				$D_{T\psi}=\iota_\psi(r(\varepsilon_\Omega))$ $\;$ $c_r=-\infty$

i are intended to represent forms of unity that *are* qualitatively distinct from each other and from the real number $|1|$—as observed by Muses [1968].)

T_X is the temporal structure that undergoes transformation in epoch X. Its dimensionality, D_{T_X}, is given by q, a value with a coefficient, i_X, whose exponent is seen to increase from phase to phase, accompanied by changes in dimensional density, c_q. The imaginary number i was identified by Muses as the "first hypernumber beyond the reals" (ϵ being the second [1985, p. 8]). In an earlier paper, Muses asserted that i furnishes "the units for the *negative* dimensions, which are deeply connected with time . . . just as the ordinary or positive dimensions are connected with space" (1968, p. 39). If this is the case, multiplying i by itself would mean transforming a nonreal number with timelike, negative, in-tensive quality ($\sqrt{-1}$), into a unit of the spacelike, positively extended, real number line, namely, -1 (since $(\sqrt{-1})^2 = -1$). Thus the operation of raising coefficient i_X to the second power can be associated with a spatialization of time, the transmutation of T_X into a primordial dimension of space, the role it assumes as F_Ψ in the initial phase of epoch Ψ.

Finally, ϵ_Ψ, enclosed by interior brackets in epoch X, is the hypernumber coefficient associated with the generation of F_Ψ. It is shown as undergoing "incubation" in X, its influence on D_{F_Ψ} not being felt until the spatialization process of epoch X has been completed.

PROSPECTIVE APPLICATION OF GENERALIZED MOEBIUS CONCEPT TO KALUZA-KLEIN COSMOGONY

This exposition has addressed what many consider to be the fundamental problem of contemporary physics: achievement of a unified understanding of nature. While affirming the widely held conviction that the key to unification could be found in a Kaluza-Klein approach entailing dimensional transformation through cosmogony, my thesis has been that such a theory cannot effectively be formulated within the traditional context of mathematical physics, since dimensional *invariance* is implicitly built into that context. To support my proposition in a constructive manner, it was not only necessary to spell out the limitations of the classical approach (section 2), but to suggest an alternative framework (section 3). The remaining space allotted does not permit a detailed application of the generalized Moebius concept to specific problems of cosmogony. Nevertheless, with the hope that the broader implications of the neo-intuitive proposal already are evident, I will proceed with a preliminary sketch, to be introduced by identifying the crux of the old intuition a final time.

Contemporary Kaluza-Klein theory relies on the notion of "symmetry-breaking." According to the version that has attracted most interest recently, the basic cosmogonic transition was from a state of "perfect symmetry" described by a ten-dimensional, "primordial" manifold, to the current 3 + 1-dimensional observable universe, with six dimensions remaining "compact." In my diagnosis of the underlying problem with this account, I have called attention to an implicit dilemma: Though theorists have found it necessary to incorporate the idea that symmetry has undergone transformation and, therefore, that a nonsymmetric process has been at work in nature, it has been at least as imperative to safeguard the classical intuition of symmetry. In fact, maintenance of this intuition has been of such high priority that not even the *suggestion* of its compromise has been permissible. Accordingly, the notion of nonsymmetry has been nipped in the bud in the very act of introducing it, not by conscious exclusion, but quite automatically. Rather than allowing the possibility that nonsymmetry could be intrinsic to the abstract framework for our experience of the universe, it is assumed to be merely an *ex*trinsic feature of an even more abstract, invariant frame of reference beyond direct experience. By thus preserving symmetry in abstraction, a gap is opened that must be "filled" by introducing parameters that are arbitrary not just in the sense that they must be adjusted in accordance with what is concretely observed (because they are not unambiguously aligned with experience) but in a stronger sense related to the fact that no direct observation can even be made that could provide the basis for such adjustments (i.e., the parameters are unalignable with immediate experience; see section 2.2). Therefore, while the abstractionist surely would claim that his formalism gives no indication of intrinsic nonsymmetry, it might be said that nonsymmetry indeed is expressed, though only implicitly and in a negative way, in the very gap that had to be opened to deny it positive expression.

Whereas conventionally formulated Kaluza-Klein cosmogony widens the hiatus between concrete physical process and its abstract dimensional representation, the neo-intuitive approach would eliminate it. To be sure, in both cases intuition is at work. But classical intuition *detaches itself* from concrete reality. This is the essence of the Cartesian "mind-body split": a categorical schism between a priori and a posteriori, normative and empirical, universal and particular, form and fact, symmetry and asymmetry, changelessness and change—with primacy always given to the first member of the pair. On the other hand, the synsymmetric intuition is not of an abstract framework with which concretely observed changes must be aligned so as to "minimize free parameters." As a *concrete* universal, the synsymmetry idea harmonizes the changeless and changing in a more intimate way. Change is dialectically internalized in synsymmetry; it arises spontaneously within the intuition itself (thereby reducing the number of "free parameters" to zero). Let us now briefly consider what the details of this intuition of self-transforming dimensionality suggest for cosmogonic process.

In the classically grounded assumption of a ten-dimensional continuum composed of two subcontinua, the idea of presymmetric primordiality is denied. Interestingly, the generalized Moebius concept also can be said to imply a ten-dimensional cosmos, with six "compact" dimensions. However, the six dimensions are compacted not as a single, juxtapositionally spacelike continuum but in a stratified manner constituted by three orders of *lesser* continuity corresponding to three past epochs of dimensional generation, each with its own order of time. Overall ten-dimensionality is obtained here by considering only the basic entities intuited by Poincaré: point, line, plane, solid. But instead of taking these fundamental structures of experience as ready-made, they are seen to develop in distinct epochs of creation. The dimensional scheme thus yielded can be expressed as follows: $(0 + 1) + (1 + 1) + (2 + 1) + (3 + 1) = 10$. Within each bracketed term, the left-hand member is the spatial component, the right-hand member the added component of time. Summation over the first three terms gives the six hidden dimensions.

The key to comprehending the proposed account of cosmogony lies in grasping the notion of *orders of primordiality* (see table 5.2). Within any epoch of dimensional generation, the highest spatial dimension may be taken as the first order of

Table 5.2

Four epochs of dimensional development

DIMENSIONALITY

EPOCH	D_0	D_1	D_2	D_3	D_4
A	$P_0\,(=F)$	$P_1\,(=T)$	P_2	P_3	P_4
B	I	$P_0\,(=F)$	$P_1\,(=T)$	P_2	P_3
Γ	I	I	$P_0\,(=F)$	$P_1\,(=T)$	P_2
Δ	I	I	I	$P_0\,(=F)$	$P_1\,(=T)$

primordiality, P_0, since this dimension is "fractional" (F), not yet fully generated (lower, integer dimensional entities, I's, are nonprimordial). The temporal dimension (T) of the given epoch would comprise the second order of primordiality, P_1, representing as it does an "in-tensive" or "implicate" (Bohm 1980) seed-structure through which an added dimension of space is seen to incubate. The developmental notion of four epochs of generation, each with its own temporal structure, requires identification of three additional orders of primordiality, as shown in table 5.2.[11]

A (Alpha) is the first, most primordial epoch, that in which the point, D_0, was the "fractional" structure that underwent "generation,"[12] with D_1 serving as time ($T = P_1$). The three additional seed-structures displayed in A respectively represent successively higher orders of primordiality or implicateness. In B, the second most primordial epoch, the point has become an integer dimensional structure, the line has been spatialized, under-going development as a fractionally extended entity, and the orders of primordiality of $D_2 - D_4$ have been reduced by one. The process continues in this manner into Δ, our own epoch, which is least primordial (relative to our experience), containing three integer dimensional structures and but two as yet undeveloped ones. Determining the space-time values for epochs $A - \Gamma$ by the same procedure used for Δ (the dimension of time is added to the dimension number of the highest spatial entity), we obtain the stratified six-dimensional structure $((0 + 1) + (1 + 1) + (2 + 1))$ regarded to exist in Δ in compact form. (Note the nonlinear quality of developmental process that is implied: Earlier epochs of development are not merely left behind but "compactly recapitulated" in the present epoch.)

Now, while the standard formulation of cosmogony is a "nonrepresentational" account that skips over concrete details (which then must be recovered in the laborious procedure of parameter reduction), cosmological events are directly embodied in the neo-intuitive concretization of mathematical physics. Thus, it becomes possible to identify in a direct manner the four epochs of dimensional unfoldment with the four forms of physical inter-action arising in the course of cosmic history. Epoch A is identified with the "generation" of the most primordial force of nature, the gravitational. During this epoch, the other three forces existed only nascently, as seed-structures. In epoch B, the strong force "froze out," in Γ, the weak force emerged, and in our current epoch, Δ, the electromagnetic force is brought into play. The incompleteness of the present epoch expressed by the fact that spatiotemporal experience is intrinsically perspectival, implies the incompleteness of the electromagnetic mode of interaction. The epistemological consequences of this are quite far reaching and will be addressed in the concluding section of this paper.[13]

5. TOWARD THE SELF-UNIFICATION OF MATHEMATICAL PHYSICS

Perspective is point of view. That is to say: Regardless of what is viewed, the perspective of an observer is the point from which the viewing is done, the subjective ground from which objects are observed. In early concrete experience, objects are seen to change in immediate relation to changes in the observer's perspective. But as experience becomes more abstract, the reality of this linkage between observer and observed, subject and object, tends to be denied; we come to say that the object only *appeared* to change, that it actually was only our perspective which changed. By thus detaching the observational framework from that which is observed, the notion of invariance or symmetry arises.[14] In its essence, classical science entails a furtherance of this detachment process.

Descartes's imposition of a "mind-body split" depended on replacing all concrete subjects with what Heidegger (1977a, p. 273) called "the special subject." While the old subject still participated in the world s/he experienced perspectivally, the Cartesian subject was a *wholly* abstract entity, a being that, by detaching itself totally from the physically embodied world, could disregard the individual's point of view entirely, view reality from a universal frame of reference that rendered all particular perspectives equivalent. In barring individual perspective, science became a "purely objective" endeavor. From its universal perspective, a "uniform space-time context" could axiomatically be projected (Heidegger 1977a, p. 268) within which all the concrete contents of experience could be treated as extrinsic, *merely* factual, that is, as having "no concealed qualities, powers, and capacities. . . [being] only what they *show* themselves as" (Heidegger 1977a, p. 268). Of course, the "uniform context" alluded to is none other than the classical *continuum*. Here we see the connection between the need to uphold the juxtapositional fullness of space and the prime directive of classical science: maintain objectivity.

Now, of the four basic forms of physical interaction, electro-magnetism is the one most directly involved in ordinary human experience. Perception in space and time is intimately associated with the behavior of light. It is by means of "photon exchange" that we encounter our environment in the characteristic perspec-tival manner. Our inability to perceive simultaneously the opposite sides of surfaces expresses the fact that a timelike interval is required for electromagnetic interaction with the other side. Of course, classical intuition does not permit us to interpret this limitation as intrinsic, as one that would attest to the incompleteness of space itself. Rather, by the "spatialization of time," space itself is assumed complete, electromagnetic inter-action being construed as a *merely* concrete event to be represented in extrinsic, topological terms.

In nineteenth-century pre-Einsteinian physics, Descartes's "special subject" appeared in the guise of the "luminiferous ether." This was the absolute frame of reference, the intrinsically changeless medium within which light was propagated, extrinsi-cally transformed. For his part, Einstein certainly was not seeking to overturn the Cartesian "subject." On the contrary, his basic aim was to *uphold* it in the face of the challenge to the ether concept mounted by the Michelson-Morley experiment. Special relativity replaced space and time with space-time expressly for the purpose of rendering all concrete perspectives (coordinate systems) abstractly equivalent. In the space-time continuum, the particular perspective one adopts does not matter; the physical world will "look the same" regardless. By thus prohibiting intrinsic interaction between particular frames of observation and the reality observed, subjectivism was avoided in relativity physics. In quantum physics, however, it has not been so easy to avoid.

It was not that classical physics ever really avoided subjec-tivism completely; it did so only in idealized approximation, by limiting itself to the macrocosmic domain where the influence of the observer could be disregarded, where the observer's electromagnetic interactions with the system observed could be taken as negligible. Just this interaction is no longer avoidable in the quantum mechanical effort to fulfill the program of tradi-tional science, to bring physics to its ultimate order of precision. The "problem of measurement" plaguing QM derives from the fact that here, the concrete perspective of the observer (required for the reading of meters, the arrangement of experimental apparatus, and so on) is dynamically, *non*negligibly linked with the system under study. This undermining of the Cartesian "special subject," this challenge to classical intuition, has weakened the hold of traditional realism, bringing many of the leading figures in quantum physics to openly proclaim the *subjective* foundation of reality (see Jahn and Dunne 1983). But note that such subjectivism does not transcend classical intuition if it does not genuinely call into question the *schism* between subject and object upon which that intuition essentially rests.

There is certainly another side to QM, one that is far from subjective. As mentioned before, the objectivity of the scientific enterprise has been preserved in QM by resorting to an order of abstraction even further removed from concrete reality than the old order. With this development, the Cartesian split is brought to its definitive expression. Subjectivity and objectivity are carried to their most extreme possibilities, where their meanings are retained in inverted form: what had been object, the objective reality that is known, becomes utterly subjective to the point of being arbitrary or absurd; what had been subject, the concrete, individual knower, and had become "special subject" under the influence of Descartes, now becomes the only *ob*jective certainty. Yet the price abstractionism must pay for this "certainty" betrays the underlying fact that the challenge to classical intuition has not been addressed effectively.

The neo-intuitive approach set forth in this exposition may be viewed as a proposal for meeting the challenge in an authentic and constructive manner, not by renouncing traditional intuition, but by attempting to surmount it from within itself. To be sure, the alternative offered is incomprehensible when one simply adheres to tradition, when the customary constraints on thinking are uncritically received. But in reflectively overcoming these constraints, classical intuition itself is understood as part of one phase (the intermediary, perspectival phase) in a threefold process of dialectical change. From the perspectival standpoint, contemporary developments may seem to suggest that only two categorically opposed alternatives are possible for science at this juncture: (1) to revert to the particular perspective of the concrete individual, a fall into subjectivism that would mean the end of science, or (2) to continue to adopt the universal perspective of higher-order abstractionism, regardless of the price that must be paid. The *neo-intuitive* alternative entails a *post*perspectival synthesis of the concrete and abstract, the particular and universal.

We are to *accept* the incompleteness of three-space, its inherently dynamic, concrete character, recognizing that it is *we ourselves*, inseparably linked with the space in which we live, married to it in electromagnetic transaction, who constitute the potential for completion. The rise of individual perspective and its progressive abstraction in Western science bear witness to the long process in which the concretely participatory aspect of our experience has been repressed. By no means has it been eliminated. And the abstractionist's dilemma, found in both quantum physics proper and modern unification theory, suggests that a lifting of the repression may be in order. This acceptance of concreteness certainly would not entail the relapse to subjectivism that would be predicted by classical thinking. Instead of abandoning the abstract universal perspective for that of the concrete particular, neo-intuitive acceptance would begin with the realization that both of these terms signify phases of development beyond the *pre*perspectival, concretely *universal* mode (the mode of undifferentiated unity). And *post*perspectival realization, far from involving a mere movement backward to concrete particularity, or further back, to the concrete universal in its primordial form, would entail moving "forward" into the very core of abstraction to discover its "inner horizon," i.e., the limitation that abstraction imposes upon *itself* through which the antinomy of the abstract and concrete is surpassed in dialectical synthesis. Thus we could consider postperspectival intuition to be the ultimate abstraction, and, at the same time, the ultimate concretion—the concrete universal "decompacted." This is the sense of concreteness in which the neo-intuitive approach would "concretize" mathematical physics. Because classical intuition requires continued repression of the feature of experience that authentically must be expressed if full-fledged unification is to be achieved, a unified account of nature must continue to elude classicism. It is this dilemma that the intuition I have offered would seek to address. Full unification would be achieved in the only way possible: By concretely including *ourselves* in that which we seek abstractly to unify.

6

The Paradox of Mind and Matter:
Utterly Different yet One and the Same (1992)

The interrelationship of mind and matter is a perennial question of philosophy. In its modern form, this problem has persisted since the time of Descartes. Despite the fact that in the twentieth century, positivistically oriented philosophers are inclined to trivialize the mind-matter (or mind-body) problem (dismissing it as merely the result of "linguistic confusion," or simply declaring it a "nonproblem"), philosophers who are still in meaningful contact with the essential questions of their tradition are likely to acknowledge that the question at hand has neither diminished in importance nor been satisfactorily resolved. In this essay, I will explore the possibility that we may meet the challenge of mind and matter through an act of overcoming our deeply engrained heritage of dualism. I intend to show that the crux of the problem before us is the loss (concealment) of wholeness and that, if wholeness is to be regained (disclosed), a fundamental confrontation with paradox is necessary.

In contemporary experience, mind is associated with our interior sense of self or ego, with that which lies under or grounds our awareness, with sub-ject, from the Latin, *subjicere*, "to throw under." Matter, on the other hand, is associated with what is manifested in the external world, with that which is sensed to lie outside us, appearing over against our subjectivity, cast before us as ob-ject, from the Latin, *objicere*, "to throw before." Now, as heirs to the post-Renaissance, Cartesian tradition, we take the division between mind and matter to be a radical one. We sharply distinguish it from, say, mere differences among objects—in their sizes or shapes, for example, or in their functions, structures, or inherent natures. For us, the mind-matter difference is *ontological*, a difference not among entities *within* a category of being but *between* such categories; not a distinction among objects but an absolute distinction between objectness and subjectness as such. In the words of philosopher W. T. Jones, who was summing up Cartesian philosophy, "body is body and mind is mind, and never the twain shall meet" (Jones 1952, p. 685). And yet, in the concrete sphere of our everyday experiences, matter and mind obviously do interact; life is somehow all of a piece. Then the mind-matter problem is just this: How can such fundamentally different entities—one inward, unextended, immaterial, the other outwardly manifested in the physical realm—interact?

To be sure, if solving the mind-matter problem meant finding a way of bringing together entities between which a radical separation has always existed, entities separated on a priori grounds, we certainly would be faced with a difficult if not impossible task. But I submit that the split between mind and matter actually has a *history*, that when seventeenth century philosopher René Descartes posited this stark schism, he was not merely identifying a state of affairs that had been eternally given in the same way; as a prime agent of a new mode of understanding, he was helping to *create* it. We should not underestimate the depth of the transformation that took place. Cartesian philosophy evidently reflected more than just an intellectual overlay on an unchanging substrate of concrete human experience. Some observers have gone so far as to suggest that Cartesianism was symptomatic of a sweeping change in consciousness, in human being itself, that occurred with the Renaissance and found expression in the art, literature, science, and philosophy of the time, as well as in ordinary life.

Philosopher Martin Heidegger is among the most perceptive contemporary thinkers to bring out the significance of the Renaissance.[1] He noted, for example, that at this time, the very meaning of the terms *subject* and *object* underwent a reversal (Heidegger 1977a). Prior to the Renaissance, it was the surrounding world, nature at large, that had been sub-ject, experienced as thrown under one, as supporting or undergirding one's sense of identity, of self (and the word *object* "denoted what one cast before himself in mere fantasy: I imagine a golden mountain" [Heidegger 1977a, p. 280]). Then self withdrew from communion with the world of concrete things, people and events, contracting to an interior perspective point, the "I" imagined to lie inside one's forehead, behind one's eyes, the ego à la Descartes.[2] Divested of subjectness, nature now became *object*. No longer lived in its immediate presence, nature became that which was to be *re*-presented within a "uniform space-time context" (Heidegger 1977a, p. 268), a framework in which all bodies could be precisely located, probed and experimented upon, submitted to mathematical analysis, and ultimately, shaped and manipulated through the enterprises of science and technology.

Others have similarly acknowledged the profound transformation of human experience occurring around the time of the Renaissance. Philosopher Owen Barfield, for example, noted that in the pre-Renaissance era "the world was more like a garment men [and women] wore about them than a stage on which they moved . . .Compared with us, they felt themselves and the objects around them and the words that expressed those objects, immersed together in something like a clear lake of . . .'meaning' " (Barfield 1988, p. 95). Philosophers David Lavery (1983) and Jean Gebser (1975) pointed to the introduction of perspective in art as indicative of this underlying change in perception: the artist now stood completely apart from others and from nature, observing them at a distance from the fixed point of reference inside his head.[3] Correlated developments in the field of written expression have been examined by language theorist Walter Ong (1977). Ong particularly stressed the far-reaching impact on consciousness of the advent of print (ca. 1450), how this new medium standardized, objectified, and spatialized language, thereby bringing to a close the more personal and participatory era of script and orality.

Note that according to Heidegger, the dynamic unity of experience existing prior to the Renaissance was not utterly obliterated but relegated to oblivion, forgotten, or repressed. It was after this act of forgetting that divided self from other, mind from matter, that we left ourselves to wonder how such "fundamentally different" entities could interact. Is it not obvious that as long as we continue to operate within the Renaissance framework of forgetfulness, the mind-matter problem cannot authentically be solved, since operating within that framework is precisely what is responsible for the problem? In the various "monisms" and "dualisms" of modern philosophy, many attempts have been made to address the mind-matter question in a meaningful fashion. But, as I have argued elsewhere (Rosen 1985), none of these efforts at solution have succeeded because modern philosophy, in essentially maintaining the Renaissance way of knowing and being, remains dualistic at bottom; implicitly if not explicitly, it presupposes an unbridgeable separation of mind and matter that obscures the concrete wholeness of experience. So it seems that if we are going to approach the problem in a genuine manner, somehow we will need to lift the repression imposed at the time of the Renaissance, recover something of the original, pre-Renaissance sense in which there was no problem, because mind and matter had not yet been starkly, categorically separated. But while wholeness needs to be retrieved, this does not mean going back to it in the simple sense of wiping the slate clean, eradicating the division between mind and matter as if it had never occurred. We could not go back in such a way, nor should we want to.

Pre-Renaissance awareness is not something to be idealized. As Gebser (1975) and Ong (1977) made particularly clear, it was less differentiated, less lucidly focused than the mode of awareness that succeeded it. The differentiation that gave rise to the mind-matter problem served the interest of creating sharper understanding, a greater capacity for reflection and intellectual achievement, and thus it helped us to fulfill our human potential. In any case, I do not believe we *could* go back.

The division of mind and matter arose through an historico-evolutional process that, in a certain important sense, is irreversible. In what sense? Here the thinking of Heidegger again becomes relevant. Heidegger was deeply concerned with history, but for him, "history is neither simply the object of written chronicle nor simply the fulfillment of human activity," but rather, "a sending," a "starting upon a way," a *"destining"* (1977c, p. 24). If history were "simply the fulfillment of human activity," it would lie at the arbitrary discretion of human design, be freely constructible under the command of the Cartesian ego, to be done or undone as suits the will. Such a history indeed would be subject to wholesale revision, be utterly reversible. For Heidegger, however, history has a weight, an indelibility beyond the caprice of human design, and therefore, it cannot merely be undone. And yet, in the first part of his comment, Heidegger states that history is not "the object of written chronicle" either, that is, not an objectively fixed truth received by us passively like the tablets handed down on Sinai, arriving ready-made from a determining source that lies outside us. History, as a "destining that starts us upon a way," is "never a fate that compels," says Heidegger (1977c, p. 25). There is a freedom in this destining, not one of "unfettered arbitrariness," but one that goes beyond "the constraint of mere laws" (Heidegger 1977c, p. 25).

Evidently, the antithetical views of history to which Heidegger objects have a common source: the transformation of subject and object. It is for the narrowed down, contracted subject, the self withdrawn from communion with nature, that self-determination means only the ego's latitude to work its will, impose its arbitrary designs on its surroundings. And when this subject fails in its attempt to shape and determine nature, it fatalistically declares that nature (now experienced as ob-ject) is determined by "fixed laws" operating impersonally, or by an omnipotent "Person." But if history is determined neither by persons nor by impersonal laws, neither by subjects (mortal or divine) nor in a merely objective manner, how is it determined?

Heidegger's answer would be that history is not determined. The syntax of saying, "history is determined by. . .," tacitly implies the operation of an agency that itself is outside or independent of history, an agency external to that upon which it acts. Again we have the split between actor and acted-upon, subject and object. Instead of saying, "history is determined by. . .," Heidegger would say it is a determin-*ing*, that is, a self-determining process, a *destining* that entails a wholeness beyond that possible when one's thinking is split into determin*er* and determin*ed*. Heidegger associates the wholeness of self-destining with *Being*—precisely that concrete, dynamic unity, that radical communion which predates the subject-object split. Here we see that although history cannot be reversed in the sense of erasing the deep division between mind and matter as if it never had occurred, as a destining, history itself operates from out of the wholeness of Being; indeed, when Heidegger speaks of the "history of Being," he is intimating that history, in its essence, *is* Being, and Being is history. Since Being is not just framed by history but is its own "silent voice," it could never have been obliterated in a bygone epoch but must always be present in an immediate way, always be "presencing," to use Heidegger's expression. It follows that Being's divisive self-forgetting must also be occurring every moment, not discernibly, from one moment to the next, but as what lies concealed "within" each moment, so to speak. Let me emphasize that the forgetting in question is no errant deviation symptomatic of a fundamental conflict or cleavage within Being. *As* radical wholeness, Being never could be at odds with itself in any simply oppositional way. The "cleavage," the "hole" within Being, is the self-concealing (w)holeness of Being itself. In a similar vein, Heidegger suggests that "self-concealing, concealment, *lethe* [i.e., forgetfulness of Being], belongs to *a-letheia* [unconcealment of Being] not just as an addition, not as shadow to light, but rather as the heart of *aletheia*" (1977b, p. 390). In sum, the unmitigated unity I am attempting to adumbrate is not merely one in which unity prevails over separation but one in which the unity of mind and matter (*aletheia*) and their separation (*lethe*) are themselves apprehended as inseparably united!

This then, is the sense in which the unity of mind and matter must be raised from oblivion, recollected, or remembered: not by an act of going back to the simpler, pre-Renaissance unity, but by grasping the more radical, all-embracing unity of pre-Renaissance unity and post-Renaissance division. It is a paradox, the term *paradox* here understood in the Zen-related sense of a wholeness so uncompromising that it confounds the dichotomies built into ordinary thinking. Perhaps the paradoxical recollection that is called for will occasion a new phase of development, a "new Renaissance," or as Heidegger has hinted, "a turning" (1977d), "another destining, yet unveiled" (1977d, p. 37).

Interestingly, according to philosopher D. M. Levin (1985), despite Heidegger's great contribution, he actually did not go far enough in overcoming the dualistic tradition of Western philosophy. He stopped short of a full recollection of Being because his holistic contemplation, profound as it was, largely was limited to the sphere of the intellect, did not encompass more concrete, embodied levels of Being. In Levin's view, for the repression on Being to be lifted in earnest, its all-encompassing wholeness must be awakened and lived at *every* level of human existence—in our bodily gestures, feelings, and concrete perceptions, for example, as well as in our conceptual activity. "Our thinking," says Levin, "will not find its way back to a more primordial presencing of Being without first 'losing itself' as a metaphysical 'thinking' and going very deeply *into* the body of experience" (1985, p. 56). Accordingly, Levin has sought to supplement the Heideggerian approach with that of Merleau-Ponty, the phenomenologist who placed primary emphasis on perception and the bodily dimensions of experience.

Now, if the recollection of Being needs to be more deeply embodied, say, through a concrete act of perception, and if said recollection entails paradox, we may ask what this "body of paradox" would "look like." The answer I offer begins with an adaptation of a comparison I have employed in several other forums [see chapters 1 and 5].

In figure 6.1a, the opposition of mind and matter is symbolically expressed in terms of visual perspective. If you were initially viewing a solid cube from the angle shown in the left-hand member of figure 6.1a, you would obtain the point of view of the right-hand member by moving 180° around the cube to the diametrically opposite side. The faces of the represented solid that are visible from the right-hand perspective are precisely those that would be concealed from the left-hand point of view, and vice versa. Of course, in our ordinary experience with perspective it is impossible to view both sides of an object simultaneously; all six faces of the cube cannot be apprehended in the same glance. Opposing faces remain mutually exclusive in our customary way of perceiving them.

As mentioned earlier, the appearance of perspective in Renaissance art was a prime indication of the loss of organic (comm)unity, its repression in forgetfulness of Being. The underlying effect of abstract perspective was to "remove the garment of reality" (Barfield 1988), to establish decisively the uniqueness and distinctness of the observer's point of view, to locate it inside his head, thereby separating this subject from the objects around him or her, setting the viewer in opposition to that which she or he was viewing. This wedge that was driven between inside and outside, subject and object, mind and matter, carried over to the observed objects themselves, giving them their appearance of depth and creating the diametrical opposition of sides illustrated in figure 6.1a.

We now consider figure 6.1b, known to modernist art and phenomenological psychology as the *Necker cube*. Inspection readily discloses that both of the perspectives shown in figure 6.1a are incorporated in the body of 6.1b. An *ambiguity* of perspective results from this (assuming we view the cube in our customary manner of viewing things). You may be perceiving the Necker cube from one point of view, say, as hovering above your line of vision, when suddenly a spontaneous shift occurs and you see it as if it lay below. But despite the fact that two distinct perspectives are established in the course of gazing at the cube, they completely overlap, are internally related, spatially integrated in a totally interdependent manner (think of what would happen to one perspective if you tried erasing the other!).

Figure 6.1

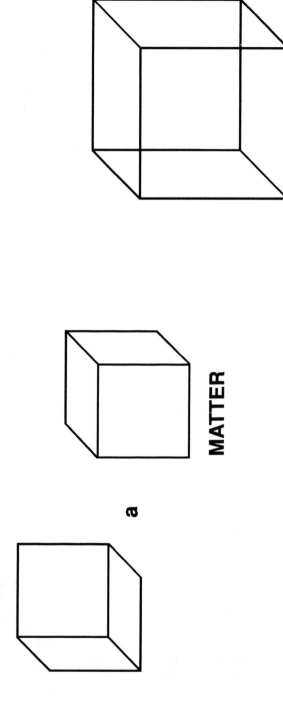

Mind and matter seen as separate perspectives (a); integration of perspectives through Necker cube (b)

As I interpret its significance, the Necker-cube relation reflects a certain contemporary advance over classical dualism, a step in the direction of unity. Time and again, I have seen just this sort of structural relationship implied (if not explicitly stated) in modern process and phenomenological theories. Blatantly dichotomous Cartesian formulations are abandoned in favor of notions of mutual dependence and reciprocity closely akin to that embodied by the cube. In two other papers (Rosen 1985 and 1984 [= *chapter 14]*), I have considered an important example of this: the process philosophy of Alfred North Whitehead (1978). In place of Descartes's sharp division of reality into mental and physical substances (*res cogitans* and *res extensa*), Whitehead posited a single entity, the "actual occasion," possessing mental and physical aspects or poles. According to philosopher David Haight (1983), Whitehead's dipolar monism is a clear improvement over the dualism of Descartes.

However, Haight goes on to observe that Whitehead did not fully liberate himself from Cartesian influence. In my own understanding, the crux of the problem lies in the Whiteheadian "rule of thumb" stated by philosopher D. R. Griffin: "subject *then* object" (Griffin 1986b; emphasis added). That is to say, while no spatial separation exists between subject and object, between the mental and physical poles of an actual occasion, there is a *temporal* separation between occasions. In the procession of events described by Whitehead, poles are taken to be related in a simply sequential, strictly successive manner. What is subject one instant, has completely "perished" by the very next, become a lifeless object, to be "prehended," mentally enfolded by a newborn subjectivity that somehow has sprung up. Subject then object, mind then matter, inside then outside—that is the rule, and by virtue of this simply successive relation, though given subjects are transformed into objects on each occasion, subjectivity and objectivity as such remain unaffected; their mutual exclusiveness and diametrical opposition are maintained. In such a conception, subject-to-object transitions can be apprehended only as utterly discontinuous events, pole-to-pole "quantum leaps" that leave us incognizant of what happens in between. Just such a transition is illustrated in the leap we take from one perspective of the Necker cube to the other, when we view the cube in the ordinary fashion.

But we actually can go a step further in our perception of the cube. Rather than allowing our glance to oscillate from one perspective to the other, we can attempt to view *both perspectives of the cube at once [as discussed in chapter 5]*. This can be accomplished by an act of mere abstraction, in which case the cube will simply flatten into an array of connected lines. On the other hand, it is possible to retain the awareness of depth; when this is achieved there is an experience of self-penetration—the form appears to go *through* itself. This mode of imaging changes the perception of the cube's faces in a critical way.

In the conventional, perspectivally polarized way of viewing the figure, when the quantum jump is made from one pole to the other, all the faces of the cube that were seen to lie "inside" presently appear on the "outside," and vice versa. But it is only at "polar extremes" that faces are perceived as *either* inside *or* outside. With the fusion of perspectives that discloses what lies *between* the poles, each face presents itself as being inside and outside *at the same time.* Therefore, the dualism of inside and out—symbolically, the dualism of subject and object, mind and matter—is surmounted in the creation of a *one*-sided experiential structure in which all six faces are simultaneously given. Simultaneously? Well, that is not exactly the case. What we actually experience with the perspectival integration of the cube is neither succession nor simultaneity alone, nor a linear combination of them, but their full-blown merger, a blending of time and space so thorough that they are no longer recognizable in their familiar, categorically dichotomized forms.

Let me underscore the fact that this self-intersecting structure does not just negate the distinction between sides (as would a perspective-flattening act of abstraction). Faces *are* inside, and yet *outside* as well. So the feature of duality is not lost; rather, a unity is gained that is deeper than that of the simply dipolar structure; without merely flattening opposition, the complete interpenetration of opposites is embodied. To be sure, this is a paradox, and if we had been limited to stating it in words alone, if we had been restricted to the abstract, intellectualized medium of language that has been dominant since the time of the Renaissance, we would have been left with naught but a sense of uncanniness, incomprehensibility. But through the concrete act of perception performed with the Necker cube, the paradox is fleshed out. It becomes a living presence, presenting itself with an immediacy that permits us tangibly to grasp it. Through this body of paradox, we can literally *see* how mind and matter could be entirely distinct and, also, one and the same. We can palpably sense what Heidegger was getting at when he hinted at the nonseparation of *lethe*—the concealment of Being that underlies the mind-matter split—and *a-letheia*, unconcealment of the primordial unity of Being.

I am convinced that the only way to "solve" the mind/matter problem is to follow the exhortation of D. M. Levin. We need to overcome the post-Renaissance, "ego-logical" (Levin 1985) framework of dualism in which the problem is rooted, and we can do this by engaging in ever more embodied recollections of Being. By re-membering, fully re-embodying that within us which is "old," we may arrive at the threshold of the *new*—a "new Renaissance;" a new, transegoic phase of human being (Washburn 1988); perhaps even an entirely new order of being.

But I suggest that if we sought to carry this out merely by circumventing the intellect and engaging the body, dualism would still be at play. In attempting to lift the repression placed upon the body in a way that surmounts mind-body dualism, the transition from the hitherto dominant intellect to the flesh-bearing body must be so thoroughly fluid that, in effect, it is a "non-transition." Differently put, we cannot simply leave the province of intellect, cross the boundary of mind and immerse ourselves in flesh, for such boundary crossing is itself an act of simple division *expressive* of the intellect. Like the proverbial Chinese finger puzzle, the more we would strive to free ourselves in this way, the more we would be entrapped.

Rather than seeking to disengage ourselves and go outside the intellect, I propose we move in the "other direction," into its depths, down through its heart and essential core, where we may discover its "*inner* boundary." "The essence of intellect is not the intellect," Heidegger would have said. Crossing this "boundary" from the intellect to its "nonintellectual" essence, what do we find on the "opposite side?" The end of opposition! Not body-as-opposed-to-mind, but bodymind. The "inner boundary" of the intellect is the "boundary" of boundedness itself. Again, we have paradox. What we are seeing is that it is *paradox* that forms the true horizon of bodymind nonduality, of the wholeness that is authentically recollective of Being. In voicing paradox, the mind is confounded and a pathway is opened to the "silent voice" of bodymind. To begin moving down that path, we must move to more embodied expressions of paradox, such as that given in the exercise with the Necker cube. The self-penetrating inside-out structure that is disclosed in the special way of viewing the cube furnishes us with a tangible insight into the "boundaryless boundary" of bodymind (or "mattermind").

In previous efforts, I have attempted to "materialize" the body of paradox in ways still more palpable [*i.e., via the Moebius surface and Klein bottle*]. But to address the challenge of mind and matter in the *most* authentic way, it seems we will need to *radically internalize* the body of paradox, embody this all-embracing wholeness with our very own bodies, through our gestures, movements, feelings, and acts of deepest intimacy. It is my conviction that a full-fledged re-membrance of Being will require us to enter the body of paradox so deeply that, in a full and literal sense, we *become* it.

PART II

The Moebius Principle in Parapsychology

INTRODUCTION

The essays in this section cover the period of my writing from 1974 to 1987 and deal with the frontier field of inquiry known as *parapsychology*. Generally defined, this controversial discipline entails the study of "paranormal" phenomena, also called "psi" or "psychic" phenomena. Included here are extrasensory perception (e.g., telepathy, clairvoyance, precognition), psychokinesis ("mind over matter"), and, most controversial, survival of biophysical death—all of which test the limits of the "possible," as defined within the framework of post-Renaissance Western rationalism.

The following articles were written over a time span largely concurrent with the essays on contemporary natural science appearing in part 1. My joint involvement with these two fields reflects the fact that the curious phenomena of parapsychology seem to parallel those of modern science, especially those of modern physics. Both areas of investigation are replete with paradoxes, and of a similar kind, enigmas calling into question our conventional thinking about space, time, and the nature of human experience. Both disciplines are thus highly problematic from the standpoint of "normal science." Why is parapsychology in particular given so little credence, surrounded by an aura of absurdity that makes it difficult for the person of "common sense" even to take it seriously? For one thing, parapsychology is not solidly ensconced within the core of science as is physics but is a fledgling venture on the periphery of established science. A more substantive reason is that the anomalies of modern physics have been treated as *remote* from everyday experience, relegated to the submicroscopic realm of the atom or to events inside "black holes." Thus, we have had the latitude to think of them in figurative terms. In parapsychology, on the other hand, the "impossible" comes uncomfortably close to home, encroaches on our lives in an immediate way so that, if we do not dismiss it, we are obliged to take it *literally*. More is said on the issue of parapsychology's legitimacy in the chapters that follow.

In chapter 7, a simple demonstration of "non-Euclidean visualization" serves to introduce the prospect that psi phenomena themselves might come to be understood in such non-Euclidean terms. This idea is substantially expanded upon in chapter 8, a freewheeling essay bringing in Einstein's theory of relativity, relating it to Eastern thought, and, in the end, expressing it by way of the Moebius principle. Chapter 9 focuses on the central importance of philosophy's mind-matter problem for the question of psychic experience, and it suggests that said experience is indicative of an epistemological crisis in contemporary thought. In chapter 10, psi is explored in relation to C. G. Jung's fourfold typology of the human *psyche* (the influence of Jung is evident in many of the chapters of this section). In the final chapter of part 2, I suggest that understanding psi may ultimately depend on superseding the pervasive dualism of Western philosophy and religion and adopting the alternative of *nondual duality*, a paradoxical notion found in the East and embodied in the Moebius principle.

7

A Case of Non-Euclidean Visualization (1974)

Have we ever been able really to deny the inherent ambiguity of our experiences so brilliantly articulated by phenomenologists like Merleau-Ponty? We must admit our past indulgence in the doctrine of realism. Viewing the world as monolithic and unchanging, we have sanctioned experiences that seem to agree with this attitude while discrediting others as being less real. But the genesis of the non-Euclidean geometries (in the nineteenth century) shook the foundations of realism. The architecture of the world no longer had to conform to a single blueprint. In the geometric pluralism that emerged, the contour of space became strictly relative to a given postulational context instead of being defined in absolute terms. Yet, though this advancement in mathematics occurred more than a hundred years ago, its implications for *visual* geometry have still not been widely accepted, especially among psychologists. Thus, in his study of the Mueller-Lyer effect, Alapack (1971) noted that psychologists continue to regard this phenomenon as an "illusion," as if the metrical equivalence of the horizontal lines is of a higher reality than their context-created difference.

One might say that the Mueller-Lyer effect is used in the present study to "take the offensive" against realism. An example presumably establishing the impossibility of non-Euclidean visualization is presented, then refuted by a simple counter-example. In the latter, the addition of Mueller-Lyer arrows can produce a visual experience that contradicts a basic postulate of the Euclidean system. The shortest path connecting parallel lines should be experienced as straight. The perception of a *curved* line as being shorter can mean nothing less than a transformation of the geodesic character of the visual space.

Hopkins (1973) attempted to support the view that non-Euclidean perception does not occur. For example, we can picture only one straight line between two points, the line we see as the shortest possible one. In figure 7.1, we cannot visualize line *a* or

129

Figure 7.1

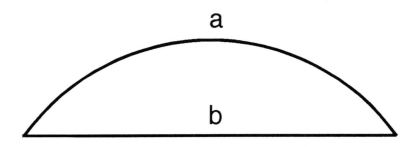

Hopkins's (1973) illustration of Euclidean visualization

any other curved line bounded by the same endpoints as being equal to or shorter than line *b.*

Hopkins admits of *non*geometric cases that display a certain perceptual ambiguity:

> There are cases in which two ways of seeing between which we can change, are available. We can picture other lines on paper—a duck/rabbit, say—first one way (as a duck), then another (as a rabbit). . . .The change can be induced by giving descriptions or by interpolating other pictures. With our way of seeing geometrically, it is not like this. We cannot first see *a* as longer than *b*, then see them as equal. Here no change occurs and no pictures or descriptions induce one. (Hopkins 1973, p. 9)

Hopkins states further (citing philosopher of science Hans Reichenbach) that if such a change in visualization did occur, it would constitute "an emancipation from Euclidean congruence" (1973, p. 9).

However, there are numerous examples of *geometric* constructions analogous to the nongeometric duck/rabbit ambiguity that illustrate the equivocal nature of perception so antithetical to the Euclidean system. The well-known Mueller-Lyer "illusion" (fig. 7.2) is especially instructive. Line *a* of figure 7.2, enclosed by acute angles, is seen as being shorter than line *b*, flanked by obtuse angles, though the lines are of equal length when measured with a ruler.

The Mueller-Lyer effect can be interpolated in an illustration similar to that given by Hopkins. In Hopkins's case (fig. 7.1), lines

Figure 7.2

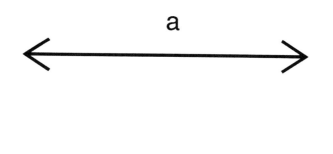

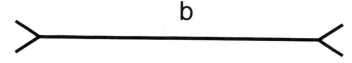

The Mueller-Lyer effect

a and *b* are bounded by the same endpoints. In figure 7.3, so as to present the Mueller-Lyer "illusion" clearly and effectively, lines *a* and *b* are bounded by separate pairs of endpoints vertically displaced from each other along parallels. A Euclidean visual

Figure 7.3

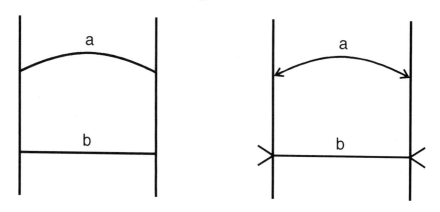

Curved line perceived as longer (left) and
as shorter (right) than straight line

geometry would require that any curve joining parallel lines be perceived as longer than a straight (perpendicular) line connecting them; this is confirmed in the left-hand member of figure 7.3. But when the Mueller-Lyer effect is introduced (fig. 7.3, right), a reversal takes place that confounds Euclidean expectation. Here, the curved line is perceived to be *shorter* than the straight line.

By showing that the Mueller-Lyer configuration produces an explicit violation of a Euclidean postulate, the first step is taken toward an extended perceptual geometry. With such a tool at our disposal, we may attempt a serious analysis of phenomena hitherto regarded as mere curiosities by the proponents of realism. One particularly intriguing possibility suggests itself in connection with the further development of the non-Euclidean approach to atypical modes of perception. A phenomenon considerably more extraordinary than the Mueller-Lyer effect—the so-called paranormal, psychic, or psi experience—might come to be understood as but another case of non-Euclidean perception. Consider the preliminary line of reasoning on which a theory of this sort could be based . . .

[At this point in the original article, C. G. Jung's approach to psi is briefly recounted, his collaboration with physicist Wolfgang Pauli noted, and it is tentatively suggested that a non-Euclidean geometry of the kind employed in Einstein's general theory of relativity might eventually make psi experiences intelligible. These closing paragraphs have been omitted from the present essay to avoid repetition, since essentially the same material appears in chapter 8 in an extended, highly elaborated form. In effect, chapter 8 is a continuation of the approach introduced in chapter 7.]

8

Toward a Representation of the "Irrepresentable" (1977)

INTRODUCTION

C. G. Jung's lifelong struggle with the problem of psi culminated in his monograph *Synchronicity: An Acausal Connecting Principle*. Here he came to the conclusion that with psychic experience we enter a realm where "space is no longer space, nor time time" (1955, p. 90) and that this domain is irrepresentable within the framework of traditional Western thinking. Yet Jung well knew that such events were not alien to the *Eastern* vision of the world. Indeed, we shall see that the synchronicity idea, only half-understood by Jung himself, was a "spice" brought back from his intuitive excursions in the Orient, beyond the limits of Western logic.

I believe that if the challenge posed by Jung can be met, a far-reaching synthesis of Eastern and Western traditions is promised. The urgency of the need somehow to represent the "irrepresentable" most recently has been articulated by Chari (1974), Pratt (1974), and Whiteman (1973). All agree that the answer lies beyond the orthodox conception of space and time (that orthodoxy including the Einstein-Minkowski worldview as well as the Newtonian).

Pratt's view of the dilemma is especially revealing. He argues that psychic happenings cannot be accommodated in an explanatory system seeking to determine the location of events in space-time. Consequently, the physicalistic space-time approach must be discarded. But if psi can be dealt with only *outside* the physical world, it cannot be dealt with at all, for we must then resign ourselves to an absolute dualism that is, in essence, an admission of our inability to explain.

Pratt's difficulty is, at its base, the same difficulty Jung had. Both writers appear to see only two alternatives: either an event can be located in conventional space-time or it must be outside space-time entirely. On the other hand, a third alternative is suggested by Whiteman. Renouncing the notion that entities must be "simply located" in space-time, he proposes that an "hierarchical," multilevel analysis be substituted for the classical, single-level conception. However, after discussing the analogy of psi to modern physics (i.e., the quantum theory) and outlining a set of axioms on which an hierarchical theory might be built, Whiteman strikes a note of pessimism. He concludes that no extension of mathematical-physical theory will accommodate psi, which is rightly regarded as *non*physical. Then is he not recommending an absolute dualism after all? Must we accept the irreconcilability of psychical and physical spheres?

Let us suppose that we could describe a unitary but open geometric structure that generates higher orders of space-time continually through a process of *self*-evolution. Might this not lead us toward a representation of the "irrepresentable"? Though we would have multiple levels of reality, the levels would not be separated by void but would grow out of each other. I will present such a geometry, bringing it to bear specifically on the phenomena of clairvoyance and precognition. At the same time, it should become evident that this approach is in fundamental harmony with the worldview of the East, an outlook always accepted by Jung, if only on an intuitive basis.

IDENTITY, CAUSALITY, AND SYNCHRONICITY

The Aristotelian law of simple self-identity is perhaps the deepest presupposition on which the classical, single-level view is based:

$$A = A \text{ and } A \neq \bar{A}.$$

The idea that bodies are simply located in space and time is a variation on this identity principle. A given body cannot be in two places at the same time, nor can different moments of time overlap one another. A more sophisticated expression of this precept is built into Einstein's special theory of relativity, where *spacelike* and *timelike* space-time intervals are specified. Essentially, two events in Einstein-Minkowski space-time are separated in a spacelike way if the space between them is great enough relative to their temporal disjunction that an observer could be present for the onset of both events only by traveling from one to the other at the velocity of light or greater. The theory assumes that no material body can travel that quickly. Events are separated by a timelike interval if the time elapsing between them is long enough relative to their spatial separation that an observer may be present for the onset of both events, but could not experience them as occurring simultaneously, since to do so again would require traveling at the prohibited velocity.

If the spacelike hiatus could be bridged, one would be violating the spatial aspect of the principle of simple location by, in effect, being in two distinct places at once. If the timelike separation were overcome, one would be in violation of the temporal aspect of the principle of simple location by encountering, as it were, two different times simultaneously (rather than in succession, as normally happens). Having seen what the violation of classical relativity implies, we are able to formulate our understanding of clairvoyance and precognition more specifically. Guided by Jung's proposition that when it comes to the synchronistic parallelism of psi experience, "space is no longer space," clairvoyance may be defined as the experience "here" of an event remote enough in space relative to its temporal displacement that the relativistic barrier of the space-time interval is surmounted; in this sense, the classical conception of space predicated on the idea of simple location is destroyed. And precognition is the experience "now" of an event remote enough in time relative to its spatial displacement that the limitation of the timelike interval is transcended, and with it, the orthodox notion of simple temporal succession.

In his struggle with the "irrepresentable," Jung framed his deliberations more in terms of the principle of *causality* than that of identity. Yet it is possible to argue that the objection to causality violation is based on essentially equivalent grounds. To appreciate this, we must understand the close relation between causality and temporal order. Russell (1925), for instance, invoked Robb's theory of time whereby we can say that event A definitely precedes event B only if it can be shown that A has some influence on B. In other words, the causal relation between A and B is suggested as the criterion of their temporal ordering: if A causes B, then A precedes B. Obviously, the statement "A causes B and precedes B" is not a contradiction. And neither is it contradictory merely to say "B precedes A," unless one would add that "A causes B," which would be tantamount to saying "A precedes B," according to Robb. The violation of causality therefore reduces to the statement "A precedes B and B precedes A." This is the same difficulty raised by Flew when he dismissed the notion of a second dimension of time as an explanation of precognition because it leads to the "contradiction" of claiming "that the *same* event occurred *both* before *and* after another event" (1959, p. 425). Here the causality-violating sequence *BAB* is, in effect, being rejected on the grounds that it violates the principle of simple location: events separated by an intervening event must be entirely distinct; they cannot be separated from each other in this way, yet *also* be the same event.

Jung's tacit acceptance of the causality/identity principle built into classical relativity led him to the judgment that phenomena like clairvoyance and precognition are irrepresentable. He thus stated that "it would be absurd to suppose that a situation which does not yet exist and will only occur in the future could transmit itself as a phenomenon of energy to a receiver in the present" (Jung 1955, p. 27). On the other hand, he was well aware of considerable empirical evidence substantiating such "impossibilities." What is more, Jung knew intuitively of their validity. Hence he was drawn to postulate his principle of "synchronicity," defined as the meaningful co-occurrence of a subjective experience and an objective physical event in a manner transcending the law of causality. And while a physical representation of synchronicity eluded him, he found an expression of his acausal principle in the Eastern way of the *Tao*.

The mantic procedure of sorting yarrow stalks—or, in modern times, tossing coins—is regarded essentially as a means of generating synchronistic occurrences. The hexagram constructed from the "chance" throw of the coins is the product of an objective physical event. Nevertheless, it is presumed to be in intimate correspondence with subjective matters of deep significance (if the individual concerned approaches the procedure with the appropriate attitude). Thus, in the *I Ching*, for each of the sixty-four possible hexagrams, far-reaching and detailed interpretations are offered involving the personal conduct of the coin tosser's life, though the outcome of the throw is a purely random event to the Western mind.

To understand fully the Taoist outlook, we must dismiss the idea that objective and subjective aspects of experience are basically separate categories of existence that are *brought* together. From the Eastern standpoint, it is quite the other way around:

The Tao begot one.
One begot two.
Two begot three.
And three begot the ten thousand things. (Lao Tsu 1972)

And Jung quoted from the teachings of Chuang-tzu:

"The sages of old," says Chuang-tzu, "took as their starting point
a state when the existence of things had not yet begun . . .The
next assumption was that though things existed they had not
yet begun to be separated. The next, that though things were
separated in a sense, affirmation and negation had not yet begun.
When affirmation and negation came into being, Tao faded. After
Tao faded, then came one-sided attachments." (Jung 1955, p.
100)

The Taoist, of course, would view the "fading" of Tao not as
ontological but only as a fading from consciousness. To the Taoist,
the "one-sided attachment" is thus both illusory and undesirable.
To achieve fulfillment, we must return to the Tao, or rather, rid
ourselves of the illusion that cosmic unity itself has ever dissolved
or that it never has existed. So the Eastern thinker *begins* with
the assumption of universal oneness. As a consequence, he has
no difficulty accepting the idea of synchronicity, though this
conception might seem entirely nonsensical to one presupposing
the fundamental disconnectedness of diverse phenomena.

The unified way of viewing the world is not peculiar to
Chinese philosophy but appears as a dominant theme in Indian
thought as well. In the *Rig Veda*, for instance, all substance and
life is seen to emerge by a process of differentiation from a unitary
field of pure potentiality—*Asat*, which is primary and undifferentiated (thus "nonexistent" in the sense of not yet being
actualized) (de Nicolas 1971). We should recognize that while Tao
and Asat express primordial universality, the particularities that
crystallize from the eternal are just as essential to the scheme
of existence. Thus Watts (1950), in his treatise on oriental metaphysics, repeatedly insisted that it is the *nature* of the infinite
to abandon itself again and again to the finite. And noting with
Watts that the infinite quality is not actually surrendered with
this abandonment, we return to Jung's exposition wherein the
Platonic notion that "all things are in all" is of central importance.
That is, when the infinite whole differentiates itself into its finite
parts (as by nature it will do), no part ever really ceases to embrace

the whole. Differentiation is to be regarded not as a process of breakage and isolation but as a means of *displaying a face or aspect*. The particularity formed by differentiation, instead of being cut off from the infinite whole, continues to be tied to the whole in multiple ways. Indeed, this theme of multiple connection arises in another approach that poses a challenge to Western rationalism. For modern phenomenological philosophy:

> *Being-in-the-world* [involves] a much richer relation than merely the spatial one of being located in the world . . .This wider kind of personal or existential "inhood" implies the whole relation of "dwelling" in a place. We are not simply located there, but are bound to it by all the ties of work, interest, affection, and so on. (Macquarrie 1968, pp. 14–15)

In Jung's essay on synchronicity, the multiple connection of the part to the whole is brought out in its fullest implication: That which appears least significant—the table crumb, the speck of dust—is nonetheless a microcosm of all that there is. By the same token, each individual mind, though finite, may embrace the whole of objective reality and thus may transcend time, space, and causality. Jung cites an additional passage from Chuang-tzu that is definite on this point:

> "Outward hearing should not penetrate further than the ear. . .thus the soul can become empty and absorb the whole world. It is Tao that fills this emptiness . . .use your inner eye [and] your inner ear to pierce the heart of things". . .This is obviously an allusion to the absolute knowledge of the unconscious [on which psychic phenomena depend], and to the presence in the microcosm of macrocosmic events. (Jung 1955, p. 100)

In sum, the Eastern system of thinking presupposes a reality in which the one becomes the many and each of the many is the one. This fundamental attitude is reflected in the *internal duality* of the yin and yang. The finite way is the way of diversity, of the opposition which is an inevitable part of life. Indeed, the yin-yang symbol of Taoism is intended as a representation of all opposites (good and bad, man and woman, dark and light, etc.). Yet because every particularity encompasses the universal, finite opposites are inextricably intertwined with *each other* as well as with the infinite.

THE INCUBATING SYNTHESIS OF *PSYCHE* AND *PHYSIS*

It is clear from Jung's own seminal statement of the synchronicity concept, and from Progoff's (1973) lucid commentary on it, that Jung's chief goal in proposing his notion was the union of *psyche* and *physis*, or "mind" and "body." It is equally apparent that he did not entirely succeed. For Jung, synchronistic phenomena like precognition and clairvoyance somehow bridge the gap between mind and matter; rather than being associated with one sphere or the other, they express the interpenetration of opposing spheres, the yin-yang interplay of *psyche* and matter as aspects of a single reality. The crux of Jung's problem was that he limited his portrayal of the material realm to the classical precepts of Einstein's special theory of relativity. (Jung's collaboration with physicist Wolfgang Pauli may have influenced him in this regard.)

In the special theory, since Einstein-Minkowski space-time maintains the flatness of old Euclidean space, the absolute future is stretched taut in a cone of light that widens upward from the point of the immediate present ("now"), while the absolute past is extended below the now in an equally rigid manner (fig. 8.1). Such a formulation renders the phenomenon of precognition quite impossible, because the temporal ordering of events inside the cones cannot be reversed. All events in the cone of the absolute future are definitely *after* the now, whereas all those in the cone of the absolute past are definitely *before* it. Hence, the future cannot be experienced now.

With respect to the spatial aspect of the light cones, their point of intersection is the immediate *here*. The spatial ordering of events prohibits the conjunction of here with *elsewhere*, the area outside the cones. Therefore, the clairvoyant experience is ruled out. Events outside the cones, being definitely elsewhere, cannot be here.

Evidently, the stretching of space-time on Minkowski's "Procrustean bed" is the means of upholding the laws of identity and causality. The relativistic separation of the fore- and after-cones actually could be annulled (Chari's [1974] remark to the contrary notwithstanding) if we were permitted to indulge in some bending of Minkowskian flat space. We would then be able to bring a point in the "absolute" future into contact with the now, though this would defy the orthodox notion of simple location in time; likewise, the here and elsewhere could be spatially united. To the

Figure 8.1

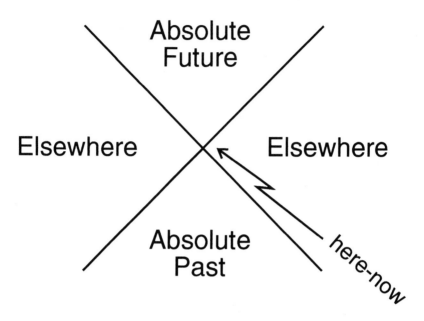

Einstein-Minkowski light cones

individual who would argue that it is not physically possible to travel at velocities greater than that of light, we would point out that this prohibition is not an ontological imperative in itself but merely a consequence of the long-cherished Aristotelian presupposition of identity/causality underlying Minkowskian geometry.

While the geometric properties of the Einstein-Minkowski light cones are consistent with Euclid's presumption of flatness, in making the transition from the special to the *general* theory of relativity, compliance with Euclid is relaxed. The implications of this for transcending identity/causality are only now beginning to be appreciated. Working with special solutions of the field equations of general relativity, mathematical physicists are starting to speak seriously of the "causally bad behavior" of certain physical systems, and even of the possibility of time travel (!) in the regions of space-time occupied by these systems (Carter 1968; Taylor 1973; Tipler 1974). Some recent efforts are directly based on Schwarzschild's original solutions of Einstein's equations generated only a year after the publication of the general theory (in 1916). Carter attributes the long delay of nearly fifty years to the fact that "many people thought that the region of space-time with the Schwarzschild horizon [the threshold beyond which time and space begin to behave peculiarly] could have no physical meaning. As a result of the interest which has developed in problems of gravitational collapse it is now generally realized that this view was mistaken" (Carter 1966, p. 1242).

A massive body in a state of gravitational collapse is popularly known as a "black hole." The ramifications of black-hole physics are startling indeed, and, in the current efforts to bring these to light, Tobin, Sarfatti, and Wolf (1975) are among the most intrepid explorers. The ordinary view of space-time as Euclidean, linear, and continuous derives from restricting our vision to the middle ranges of the scale of magnitude. However, at scale extremes, linearity breaks down and space-time displays marked curvature eventuating in the formation of black holes. The distortion of space-time thus occurs in relation to not only bodies of cosmic proportion (e.g., stars in a state of gravitational collapse) but microcosmic entities as well. Sarfatti therefore can speak of both the massive black holes of astrophysics and "miniblackholes" (Tobin, Sarfatti, and Wolf 1975, p. 127), which—in relation to physicist John Wheeler's (1962) conception of the fine grain of space—are the mouths of "wormholes."

What do cosmic black holes and miniscule wormholes (or miniblackholes) have in common? Euclidean space-time is *simply connected.* This is but another way of saying that the principle of simple location is upheld: different points in space-time cannot overlap. But cosmic or microcosmic black holes, in distorting Euclidean space-time, transform the simply connected continuum into one that is *multiply* connected. This means that regardless of the degree of separation of points in conventional space-time, they may be geometrically identified so that being at one implies instantaneous presence at the other. It is as though a "shortcut" has been created between distant regions of the manifold, but to take the abbreviated route, one must leave the Euclidean realm.

Wheeler (1962) offers an elementary illustration in which we are asked to punch two holes in a sheet of paper, one in the upper portion, the other in the lower. The distance between these points represents their separation in conventional space-time. Now we bend the paper so that the holes are brought together. The two holes are to be thought of as the mouths of a wormhole that connects remote regions of space-time. While the disjunction of points remains as great as before for the traditional journey along the surface, the traveler may reach his objective immediately by going *through* the surface via the wormhole. In this way, if the conventional space-time separation is spacelike, distant points may be superimposed. And if the Euclidean separation is timelike, definitely successive moments may be rendered simultaneous. Hence, the classical law of simple location is superseded.

Were the geometry of sensory perception to become non-Euclidean and develop multiple connectedness, it would be easy to see the possibility of clairvoyance and precognition. But events taking place in Wheeler's submicrocosmic domain or in regions of the universe where stars are in a state of gravitational collapse seem far removed from the direct experiences of organisms in the middle scale of nature. Here is Sarfatti's contribution: he proposes an entire *hierarchy* of space-time continua or gravitational fields that bridges the chasm between the remote extremes of magnitude. He is thus able to speak of a *biogravitational* field that is well within the reach of human experience:

> There might indeed be a hierarchy of finite-range gravitational fields. . .of which the conjectured biogravitational field is of the most immediate interest. The bubbles [wormhole mouths] in the biogravitational field are quite large. . .and can, therefore, directly affect our sensory perception. Indeed, I suspect that many paranormal phenomena may be explicable in terms of biogravitational wormholes. . .providing direct multidimensional links between living systems. (Tobin, Sarfatti, and Wolf 1975, p. 136)

And, more specifically:

> [An organism] in a high state of consciousness can artificially create blackholes. . .in his local biogravitational field. This would produce. . .very large distortions in his local subjective space-time environment [so that] working within his local light cone [he] "sees" into the probable future. . .This is the likely biogravitational mechanism for precognition. . .The participator can also make use of. . .modes of the biogravitational field that act outside the light cone. In this way he can. . .[receive] at a distance actions of the type reported in experiences of. . . clairvoyance. (Tobin, Sarfatti, and Wolf 1975, p. 153)

"ALL IS CONSCIOUSNESS"

A major theme of Tobin, Sarfatti, and Wolf's exposition is that "all is consciousness." At every level of organization in the hierarchy of space-time domains, singularities or holes develop at the fringes, destroying the continuity that prevailed in the middle regions. Consciousness is identified as the "hidden variable" that creates the holes and then fills them, restoring continuity. In the process, the next level of the hierarchy is produced. Thus the secret thread with which plural realities are sewn together is consciousness. An "unmanifest universe" exists, as philosopher Oliver Reiser (1966) would say, a "cosmic field" guiding the hierarchical unfoldment of all subfields, and it is *pure consciousness*. Therefore, in his opening statement, Sarfatti offers "the idea that consciousness is at the root of the material universe" (Tobin, Sarfatti, and Wolf 1975 p. 126), and later, he is prompted to say that his explanation of psychic phenomena requires accepting the "additional postulate" that the distortion of the biogravitational field associated with precognition and clairvoyance is created by consciousness.

Let us examine more closely the relation between "material" space-time and the field of cosmic consciousness or "subether" (to again quote Reiser). Carrying Wheeler's illustration a step further, imagine space-time to be riddled with wormholes or miniblackholes, so that every point is immediately accessible to every other point. What is the role of cosmic consciousness here?

In the above-mentioned paper-bending exercise suggested by Wheeler, the union of remote points is obtained by movement through a third, "higher" dimension. That is, in folding the paper, we make use of the dimension perpendicular to the two-dimensional surface. (Of course, this is a simplified, lower-dimensional analogy of our actual situation, that involving the need to "fold" our *four*-dimensional space-time.) In thus identifying "distant" points, superimposing them on each other, the "shortcut" is taken that allows us to be in two separate places *here*, or in the future *now*. Yet there is an obvious limitation to Wheeler's analogy. For, while folding the sheet of paper helps us to see how remote points in space may be brought together, such an action has an arbitrary quality, inasmuch as it is external to the space acted upon, imposed upon it from without. An adequate model of psi clearly would require an *internalization* of the bending of space.

In attempting to address this problem, let us begin by clarifying the nature of the separation that psi would overcome. The proposition we are considering is that psi experience depends not on transmitting information across a given interval of space-time but on the "folding" of space-time through a higher dimension. Therefore, rather than representing the separation of points in terms of distance along a surface, it appears more appropriate to regard the points as positioned on *opposite sides* of the surface. Now the linkage of separate points can be established *only* by making use of the dimension at right angles to the surface. And the question now becomes one of how points on opposite sides can come together in a nonarbitrary, intrinsically natural way. (In terms of Wheeler's paper-bending exercise, this is equivalent to asking how the paper could be folded without calling upon an external agency to do the folding, how the paper might fold *itself*, as it were; note that when the paper is folded we are indeed bringing its opposite side into play; in the unfolded surface, only one side is visible.)

Critically important to the question at hand is the geometric structure of the space in which points are opposed. If the topology of the continuum is the conventional, *two-sided* one (as is the ordinary sheet of paper in Wheeler's exercise), opposing sides in fact would be well insulated from one another; there would be no natural, continuous path that would connect them. In that case, a journey from one side of the continuum to the other would be physically prohibited; the far side would be physically inaccessible to a traveler on the near side. In the absence of a continuous path from side to side, passage between sides would necessitate a *rupture* in the continuum. And, having broken the continuity of space-time, the hole would have to be regarded as *outside* the material world, as though "infinitely distant."[1] But such holes *are* created, we are told! By entering them, clairvoyant and precognitive experiences may be had.

The difficulty arises in supposing that holes must be *literally* produced in the physical continuum. Consequently, they would have to be *non*physical, leading us to postulate an ethereal medium, an unmanifest field of pure consciousness existing *in addition to* the physical world and providing hidden continuity for the quantum leaps from point to point in space-time.

Do we have any choice? The idea of multiple connectedness, though mathematically legitimate, seems to suggest a physical impossibility and so to warrant the dualistic supposition of a nonphysical continuum. Yet a mathematician might insist that the notion of "wormholes" or "black holes" in space-time is just an analogy, that the mathematics of multiple connectedness simply generates the *effects* of holes and that no actual holes need exist in the fabric of reality. If the mathematician is correct, the task then becomes this: to bring the mathematical description of the world into line with its physical description, to show that higher-order mathematical continuity can be given concrete, physical expression. By achieving such a goal, the need to propose a nonphysical medium for the hierarchy of physical realities would be obviated. The purely psychical medium would become unified with the mediated, the physical. With *psyche* and *physis* thus synthesized, we would be truly entitled to say that "all is consciousness" but could just as readily say "all is *physis*," for the fusion of mind and matter envisioned by Jung would have come to pass.

[Today the foregoing argument seems clearly reductive to me. I certainly still believe that discontinuity is not merely absolute. But today I would not seek to render it utterly relative. Rather than claiming that no actual holes in space-time occur—thus reducing discontinuity to continuity, psyche to physis—I would propose that holes both do *and* do not *occur. It is my present understanding that a genuine integration of psyche and physis requires the paradoxical integration of discontinuity and continuity themselves, not the reduction of one to the other. Note that in the next section, anticipating that I might be charged with reductionism, I sought to defend myself. Today I am not convinced by this 1977 defense.]*

"HOLES WITHOUT HOLES"[2]

Though the astonishing implications of general relativity were largely ignored until recently, we must now admit that one mathematical physicist *was* quick on the uptake. As far back as 1922, Hermann Weyl offered a tantalizing clue to the potential ramifications of the non-Euclidean approach. He speculated: "it is certainly possible in the present case [i.e., the case of non-Euclidean topology as opposed to the geometry of the special theory] for the [light] cone of the active future to overlap with that of the passive past; so that, in principle, it is possible to experience events now that will in part be an effect of my future resolves and actions" (1922, p. 274).

Sarfatti also cited Weyl's enigmatic proclamation. But he did not mention that the suggestion proposed by Weyl was made after considering a topology with the surprising property of *one-sidedness*. The *band of Moebius* is the most basic example of this sort of structure. . . *[What followed in the original essay was a description of the properties of the Moebius strip, as in chapter 1.]*

Now, we have seen that the problem of perforated space-time is related to the assumption of two-sidedness. Disjunctions in conventional space-time are clairvoyantly or precognitively surmounted, in effect, by the action of linking up with points on the "other side" of the continuum. But we have found that when the continuum is presumed to be simply two-sided, the movement in the higher direction (perpendicular to orthodox space-time) would have to carry the traveler *outside* of the material universe, *through a hole.*

Then why not consider the space-time continuum to be topologically *one*-sided like the Moebius strip?[3] This would make synchronistic linkages quite continuous with physical reality. It is true that *locally*, that is, at a cross-section of the Moebius band, the band is indistinguishable from an ordinary cylindrical ring whose two sides are indeed entirely segregated from each other *[see fig. 1.2]*. Thus, viewed cross-sectionally, the Moebius appears to maintain the separation between the here-now point and points that are spatially and temporally remote. But as soon as our scope of awareness is broadened beyond the local cross-section, we see how one side of the Moebius band commences its hyperdimensional, "yin-yang" twist, encompassing the other side and, thus, the whole. By supposing another side of conventional space-time, the effect of a hole is created; but when the continuum is regarded as topologically one-sided, physical continuity is recovered. In this way, conceiving wormholes or black holes to actually be Moebius structures, we achieve an ultimate Wheelerian ideal: "holes without holes!" Or, we could alternately say that the Moebius strip models the interior region of Wheeler's wormhole, for, in his illustration, the traveler is simply seen to enter one wormhole mouth and come out of the other. However, through Moebius transport, we witness the *continuous* pathway taken in the instantaneous concatenation of distant points.

To avoid the possibility of being misunderstood, let me stop to emphasize that the restoration of physical continuity by no means implies the *reduction* of psychical phenomena. The reductionist seeks not to explain such occurrences but to *explain them away*, as Mundle (1973) recommended we do in the case of precognition. However, rather than reduce the psychical to a lesser reality status, I am proposing here an accommodative *expansion of the physical*. Thus, the restoration of physical continuity is viewed not as returning to the orthodox space-time framework but as establishing a higher order of continuity. Actually, in what we are about to see, if any "reducing" is to take place it will be a reduction of the *physical*, at least in the old, mechanistic sense of that word. Physical reality is about to take on a decidedly organic complexion. Twisting, growing, changing from within, it will be seen as infused with the glow of living consciousness.

To begin, let us recapitulate the notion that multiple connectedness leads directly to a pluralistic conception of reality. A multiply connected topology promotes an expansion of the here-now point by bringing it into immediate contact with all other points at a given level of space-time, creating the effect of an all-encompassing, elevated sense of here and now. If this broadened awareness is imagined to be embedded in its own, higher realm of multiply-connected space-time, a whole hierarchy of domains suggests itself. Sarfatti thus speaks of an ascending order of gravitational fields for each level of organization of matter. Whiteman's viewpoint, stated at the outset, is in harmony: we can comprehend the phenomena of psi only by abandoning the single-level outlook (where simple location is the only mode of being) for an hierarchical world paradigm. Other parascience writers have expressed equally consonant positions (e.g. Dunne 1938; Reiser 1966). However, it seems to me that the unity underlying the diversity cannot clearly be appreciated from these previous visions precisely because they did not address themselves to the limitations of a two-sided topology.

No void separating the strata of reality can be tolerated if unity is to be as fundamental as multiplicity. Even filling the void with an ethereal mediating substance is inadequate, for it gives the impression that physical evolution is guided simply by a *non*physical—therefore, external—agent of change. The notion of a nonmaterial "cosmic midwife" thus functions to *preserve* the old, mechanistic conception of matter. What is required instead is a *synthesis* of the physical and the nonphysical wherein reality, working through matter, continuously transforms *itself*, and, in so doing, takes on organic or "spiritual" properties. Reiser's hope of a "spiritual materialism" would then be realized. The one-sided strip of Moebius, when properly generalized, can lead us toward this goal . . .

[*In the final section of this essay, the "synsymmetric" Moebius-Klein generality is set forth. . .*]

Higher dimensions emerge in this manner, through the ceaseless dialectic of symmetry and asymmetry elsewhere referred to as "synsymmetry" [*see chapters 2 and 5*]. Here is the hierarchical conception of reality envisaged by Sarfatti, Whiteman, Reiser, and many others. It is a view of existence that transcends the classical dictum of simple location, thus bringing the phenomena of psi within the purview of natural description, rendering them *representable*. In synsymmetric reality, not a hole can be found (though we might say that the effects of holes are created when the asymmetry of a given space-time domain is recognized). It is therefore unnecessary to attempt to imagine the unimaginable—we need not think of the transition from one level of reality to another as a leap through void nor even postulate a field of pure consciousness as the cosmic facilitator of this passage between physical strata. Consciousness and physical substance are thoroughly integrated in the Moebius-Klein geometry. The physical world, being thus embodied, is seen to be a living, evolving organism, one that possesses mind. *Physis* and *psyche*, in their old dualistic connotations, do not survive this metamorphosis.

Hence—in the "place" where "East meets West"—we discover the synchronistic world: a world in which *psyche* and *physis* interpenetrate, one in which clairvoyance and precognition are comprehended as natural occurrences in that they express the ongoing process of self-transformation, the unbroken growth from within of higher orders of space and time.

9

Psi Modeling and the Psychophysical Problem (1983)

While molehills of data from spontaneous case studies and laboratory investigations of psi have risen into mountains, the phenomena of extrasensory perception (ESP) and psychokinesis (PK) have become no less anomalous. Familiarity does not guarantee meaning, and now, more than ever, meaning is precisely what the embryonic science of psi seems to require. Without meaning, an orienting context that would convincingly ground the data, the "mountains" could never be more than castles in the sky.

Not that we have not tried. Interesting theories and models of psi have been offered all along and continue to be developed by ingenious minds. It is my contention, however, that in our fascination with the virgin territory to be mapped, we have neglected the question of the mapping procedure. Granting the fact that new maps are required when exploring new territory, I propose that in exploring the "terrain" of psi, a new *style* of mapping is necessary. I am suggesting, in other words, that far-reaching epistemological consequences attend the attempt to model psi and that by and large these have gone unperceived. Therefore, whatever the particular model of parascientific phenomena has been, it has been cast in the epistemological framework of old science. I believe and shall attempt to show that to do full justice to the *para*scientific, a *para*epistemology must be created, so to speak.

At this point, an objection might be raised. Have not some of the most advanced models of psi been founded on the epistemology of the new science, particularly, modern physics? The rub here is that modern physics, far from having its own house in order, is currently in a deep state of crisis, as I *[e.g., in chapter 4]* and others (e.g., Bohm 1980) have indicated. Contemporary physicists, in attempting to deal with the aberrations of their field, have not yet offered a new and effective epistemology for doing so.

In the pages that follow, the anomalies of psi and modern physics that have come into prominence in our present century will be viewed as two horns of a dilemma centering around the general problem of *psychophysical interaction*. As I see it, this is the fundamental problem to be addressed if we are to meet the epistemological challenge now before us.

THE PROBLEM OF PSYCHOPHYSICAL INTERACTION

The question of *psyche, physis*, and their interaction is as ageless as it is profound. But the statement of this issue that is most familiar to *us*, as heirs to the post-Renaissance Western scientific tradition, was given by René Descartes some 350 years ago. *Physis*, or "body," is that which is extended and divisible; *psyche*, or "mind," is that which is unextended and indivisible. Descartes proclaimed his conviction that somehow they interact. Yet the effect of his treatment was to drive a wedge between them that would make it impossible to understand how they could do so.

The Cartesian distinction between mind and body is an absolute one. Descartes emphasized that "there is a great difference between mind and body, inasmuch as body is by nature always divisible, and the mind is entirely indivisible" (quoted in Jones 1952, p. 684). Philosopher C. T. Jones summarized the Cartesian position in the following way:

> Minds . . . are completely free and spontaneous. Being unextended, none of the laws of motion apply to them. Minds contemplate eternal truths, and enjoy and cultivate values. Bodies, on the other hand, being parts of the material plenum, are machines, and their behavior is completely predictable in accordance with the laws of motion . . . mind and body are absolutely distinct . . . Body is body and mind is mind, and never the twain shall meet. (Jones 1952, p. 685)

Divorcing *psyche* from *physis* as starkly as he did, it is evident that Descartes's chief concern was *not* with an explanation of their interaction. Descartes was a man with divided loyalties; he was a faithful follower of the Church and, at the same time, a strong advocate of what was then the new science. His earnest desire to effect a compromise led him to reason that "if minds and bodies are completely different kinds of things, and if the truths about each follow from the distinct nature of each, it is completely impossible for the science of minds and the science of bodies to contradict each other" (Jones 1952, p. 676). In questioning Descartes's view, Jones commented that "as long as it was possible to *forget about* interaction, the Cartesian compromise had a specious plausibility" (1952, p. 685; emphasis added). The converse of Jones's assertion is that as long as we can forget about the radical disjunction of mind and matter introduced by Descartes, psychophysical interaction can have a specious plausibility—and not a moment longer.

Given the orientation emerging in the Age of Reason, the Cartesian severance could hardly have served the purpose of maintaining *psyche* and *physis* as separate but equal domains. The unextended *psyche* was expelled from extended *physis* so as to "purify" *physis*; the status of *psyche* as such was of lesser interest. In this regard, psychologist Marie-Louise von Franz, from her Jungian perspective, spoke of the seventeenth-century stiffening of scientific rationalism abetted by Descartes and asked: "Where was the *psyche* in all this?" Her answer: "Its existence was simply denied" (1975, p. 40).

With the purging of *psyche* from *physis*, a "spirit of mechanism" (Schultz 1981, p. 16) was established through which the program of achieving mastery in the material universe could advance. And this is what we have witnessed and been party to for the past three hundred years: a scientific, technological, and industrial thrust that has brought us an unprecedented measure of external control. Throughout the period, the mainstream philosophical underpinning has been variously expressed, but in one way or another the idea of unextended *psyche* has been excluded. Thus, it has been spuriously "reduced" to the extended, as in many forms of so-called psychophysical parallelism and brain-mind "identity theory," or just declared nonexistent, as in varieties of nineteenth- and twentieth-century positivism. In this essentially monistic approach to the problem of psychophysical interaction, the need to explain interaction is obviated, since one of the members in the pair that would interact has, in effect, been eliminated.

THE CHALLENGE TO CLASSICAL EPISTEMOLOGY

I intend to show in due course that the outlook of simple monism is not equal to the challenge of the twentieth century. By way of preparation, I would like to shift gears, recasting the psychophysical problem in terms of a critical issue in the epistemology of mathematical physics.

To reiterate once again Descartes's basic definitions, we have *physis* as that which is extended and completely divisible, *psyche* as the unextended and indivisible. Mathematically, extendedness and infinite divisibility are the attributes of a *simply continuous space*, that is, a space that is infinitely and homogeneously dense, all "filled in," as it were, so that no region is thinner than any other, and there are certainly no "holes."

The assumption of mathematical continuity is the cornerstone of the entire edifice of Western scientific epistemology. All classical analysis of the physical world depends on it, and Leibniz and Newton, using Descartes's system of coordinates, built in this assumption when devising the calculus, the key tool in the analytic repertoire of traditional science.

As long as the continuity assumption worked, as long as the unextended, the indivisibly discrete, that is, the element of *psyche*, could be avoided, equations could be written that were fully descriptive of the dynamic pattern of events in the external world. Description led to determinate prediction (error terms existed to be sure, but they could be taken as negligible), and prediction led to control. It was by this method that the technological triumphs of the post-Renaissance era were accomplished. In a time of such fruition, there was no real reason for science to concern itself seriously with the problem of *psyche*, and the official philosophy of science followed suit.

Then, one hundred years ago, there came a turning point. In 1881, the famous experiment of A. A. Michelson on the "aether wind" set the stage for the rise of modern physics. One year later, the systematic study of "paranormal" phenomena began in London, with the establishment of the Society for Psychical Research. My contention is that these two realms of exploration are the horns of a dilemma forcing serious reconsideration of the psychophysical problem, rejuxtaposing what Descartes had rent asunder. Each in its own way introduces an "error term" into the classical analysis of the material universe that *cannot* be neglected, a "hole" that is unfillable in classical terms. Each raises the spectre of unextended *psyche* in the very midst of *physis*.

THE CHALLENGE OF MODERN PHYSICS

The transition to modern physics was made when science ventured beyond the familiar middle ranges of observation and measurement to probe scale extremes. Thus, in the mind of Albert Einstein, the Michelson experiment (later repeated with Morley) raised questions about the properties of space and time at ultra-high velocities (those approaching the velocity of light), leading him to postulate his special theory of relativity. Einstein expanded on this in his general theory, which sought to incorporate the effect of gravitational mass in the space-time account: a significant "curvature" or "warping" of space-time is associated with cosmically scaled masses. The other chief concern of modern physics is with the world of the very small, the microcosm, the domain of quantum physics.

What did science encounter in attempting to extend its program to the extremes of its scales? Could the analytic enterprise proceed as before, the same level of unequivocal mastery being attained? Quite to the contrary. Particularly in quantum physics, and in the ultimate consequences of general relativity, "infinities" or "holes" arose, "singularities" turned up in equations fashioned classically to describe the workings of physical nature. In short, the root presupposition of the analytic method was called into question; the notion of simple mathematical continuity was threatened with eclipse. Philosopher Milic Capek remarked accordingly that "the concepts of spatial and temporal continuity are hardly adequate tools for dealing with the microphysical reality" (1961, p. 238), and physicist Steven Bardwell asserted that the "failure to deal with . . . built-in discreteness is the source of the manifold contradictions in quantum field theory" (1977, p. 6). Science was and continues to be confronted with fundamental discreteness, the very quality of indivisibility or unextendedness that Descartes attributed to mind. Little wonder that theoretical physicist Arthur Eddington should characterize "the stuff of the world" as "mind stuff" (quoted in Koestler 1973, p. 58), or that numerous accounts should be written paralleling the decidedly nonclassical phenomena of modern physics with Eastern doctrines of *psyche* (e.g., Capra 1975; Zukav 1979), doctrines to which C. G. Jung was drawn in collaboration with physicist Wolfgang Pauli (Jung 1955).

Elsewhere I have offered a preliminary model for the particular pattern through which discreteness or discontinuity is manifested in modern physics *[see chapter 4]*. Giving only the general outline of my proposal here, it is that (1) discreteness arises in two closely related, complementary ways ("gravitational" and "inertial") at both the cosmic and the microcosmic ends of the scale of magnitude and that (2) these complementarity pairs are themselves complementary aspects of the same underlying reality, because the radically nonclassical state of affairs obtaining at scale extremes entails a *collapse* of scale whereby microcosm and cosmos become inherently associated. As I view the upshot of my earlier examination for the present essay, in the final analysis, the interrelated expressions of discreteness found at the frontiers of physics are fundamentally linked to the *psyche*.

Let us now turn to the other horn of the psychophysical dilemma facing science.

THE CHALLENGE OF PSI

Acknowledging the great intellectual discomfort the subject of ESP caused him, renowned information theorist Warren Weaver declared "that I cannot explain away Professor Rhine's evidence, and that I also cannot accept his interpretation." In like manner, behaviorist Donald Hebb admitted that his rejection of the strong evidence for telepathy "because the idea does not make sense" was "in the literal sense just prejudice." Another scientist wrote, "not a thousand experiments with ten million trials and by a hundred separate investigators" would persuade him to accept ESP. These reactions (reported by Koestler 1973, p. 19) give an indication of the still-prevalent attitude of the scientific establishment toward the phenomena of psi. Why do emotions run so high?

The concept of mathematical continuity implies the idea of simple location. In the experience of extended space, here is here, not elsewhere; with regard to time, now is now, not otherwhen. But in imaginative thought, and more so, in fantasy and dream, space-time codes for "proper behavior" are violated with impunity. The experient may find him- or herself at distant locales in the same instant; past, present, and future may blend "perversely" in an indiscriminate now. The principle of spatiotemporal continuity—a classically unimpeachable law in the realm of extended *physis*—is blithely disregarded. Of course, this is the province of the *un*extended or discrete, the province of *psyche*. And as long as the phenomena of *psyche* remain in their province, there is no cause for alarm. With the Cartesian compromise kept intact, science may choose simply to ignore the inner happenings of mind, or to study them as *epi*phenomena, completely confident that when they *are* externalized, they will "behave themselves," register properly as measurable neurophysiological events, in full conformance to the space-time proscriptions of *physis*.

But suppose there were an externalization of *psyche*, a manifestation of it in the extended world that did *not* "behave itself," did not lose its "perverse," unextended character. If validated, such an event indeed would be disturbing to the loyal followers of an epistemological tradition entirely dependent on the Cartesian purging of *psyche* from *physis*. In fact, it would be unacceptable. The psi phenomenon appears to be such an event.

In extrasensory perception, a direct and veridical awareness of external reality is claimed. On this score, ESP would be more akin to sensory perception than to the various forms of thought and imagination. However, in sensory perception, known receptors respond reliably and measurably to systematically manipulable physical energy forms. The parapsychologist Rex Stanford noted by way of contrast that with ESP, "We have no knowledge . . . of either a specific form of energy which might be involved . . . or of an organ which might be the receptor for this information-carrying energy" (1978, pp. 198–99). Then do we not have unextendedness here amidst the extended? Not necessarily.

It is true that the energy channel for psi transmission has not yet been detected, but this does not necessarily mean that a channel does not exist. However, if the supposition of a hidden continuum were correct, the data collected should at least reflect the *functional* presence of spatiotemporal continuity; they should indicate that basic expectations for space-time transmission from one locality to another have been met (such as the attenuation of the signal in direct proportion to distance traversed). But the bulk of the data give no such indication, as most parapsychologists acknowledge.

Nevertheless, an objection might still be raised by the proponents of the Cartesian compromise. Since the data of parapsychology are not totally consistent and unequivocal in their indifference to spatiotemporal law (e.g., Osis 1965), perhaps we should not be so quick to discard a principle so fundamental to our thinking. After all, data patterns can always be questioned in empirical science, especially in an embryonic science like parapsychology.

Yet, while the radical implications of psi might be resisted successfully at the empirical level, one should not fare as well at the *theoretical*. A unified model of psi must encompass *all* major forms of paranormal phenomena. With regard to ESP, let us grant for the sake of argument that some forms could occur by transmission, at least conceivably. Could *all*?

The intent of Einstein's space-time theory of relativity was not really to abolish the essential distinction between space and time. As physicist George Gamow plainly put it: "Whereas you can move forward, or to the right, or upward in space, and then come back again, you cannot come back in time, which drives you forcibly from the past into the future" (1961, p. 71). Continuing in his colorful vein, Gamow asserted that "nobody—not even the great Einstein—can, by covering a yardstick with a piece of cloth, waving a wand, and using some such magic phrase as: 'pee-times-co-que-time-contra-variant-tensor,' turn it into a brand new glittering alarm clock!" (1961, p. 82). Gamow would certainly agree that the reverse act of prestidigitation should be equally unperformable: clocks cannot be changed into yardsticks, time not transformed into space *[not as these are* classically *conceived, I would add]*. But this is precisely what would be required to explain temporal ESP in terms of transmission.

What we are conceding is that a telepathic or clairvoyant message could be borne on some wave or particle and transmitted across space with the passage of time. But precognitive information would need to be transmitted backward "across" *time*, and this would require a far more prodigious bit of theoretical legerdemain in the accounting. Not that such explanatory feats have gone unattempted. Expressing himself in the formalism of quantum physics, theorist Adrian Dobbs (1965) hypothesized a continuous "time" space through which messages borne on "psitrons" could be "sent" from the future into the present. Since this whole style of modeling will soon be the central focus of my attention, I will refrain from commenting further on it here. Let me just note that although many advocates of the Cartesian compromise have managed to reconcile their faith with the idea of ESP in general, precognitive ESP has proven too hard to take. One philosopher's response is especially indicative of the depth of the conflict. C. W. Mundle (1973) actually urged that we cease our attempt to explain precognition and instead seek to explain it away!

On the other hand, there are thinkers less staunchly committed to the Cartesian compromise (e.g., Brier 1976; Pratt

1974) who recognize that precognition in particular and psi phenomena in general assert themselves within *physis* too convincingly to be explained away, yet are not of *physis* and so cannot effectively be explained by any physicalistic conceptualization of space and time. This, then, is the challenge of psi. It is also the challenge posed by modern physics, as we have seen. In both cases we find the unmistakable trace of the discrete within the continuous, the unextended within the extended, of *psyche* within *physis*. Both call us back to an issue that has been sidestepped for centuries but never really solved—the problem of psychophysical interaction.

THE PERSISTENCE OF THE CLASSICAL EPISTEMOLOGY AND ITS CONSEQUENCES

Given the common non-Cartesian character of parapsychological and modern physical phenomena, and given the fact that modern physicists have been attempting to deal systematically with the paradoxes in their field for many decades, it is not surprising that those with a serious interest in psi should consider adopting the approach of the established discipline, as indeed Adrian Dobbs and many others have done. Accounts and reviews of these efforts appear in the literature. In this presentation, concern will center not on details of differences among particular models but on the style of modeling tacitly assumed appropriate and shared by all: the method of *mathematico-logical formalism.*

Does mathematical formalism provide an effective response to the non-Cartesian challenge? Today formalism is the dominant approach among mathematicians. Theoretical physicists are trained in this method, and it is generally taken for granted that this is the only correct way to deal with the problems in their field. To gain a psychophysical perspective on formalism let us examine its historical roots.

Mathematician Morris Kline (1980) gives a penetrating diagnosis of the current condition of his field in the book *Mathematics: The Loss of Certainty.* In tracing post-Renaissance origins, when certainty appeared to be assured, Kline begins as we have begun, with Descartes. For Descartes, mathematical certainty was based on the idea of intuition:

> By intuition I mean . . . the conception of an attentive mind so distinct and so clear that no doubt remains to it with regard to

that which it comprehends. . .Thus everyone can see by intuition that he exists, that he thinks, that a triangle is bounded by only three lines, a sphere by a single surface, and so on. (Descartes quoted in Kline 1980, p. 230)

As Kline observed, this notion of intuition was formulated more systematically by philosopher Immanuel Kant, who held that particular perceptions are contingent, always subject to change, but that all perceptual awareness is organized in terms of a fundamental, immutable intuition of space and time. In the words of Fuller and McMurrin, Kant took the position that "no matter what our sense-experience was like, it would necessarily be smeared over *space* and drawn out in *time*" (1957, pt. 2, p. 220). Thus, spatial extendedness and temporal sequentiality were invested with a priori status, viewed as unqualifiable givens; these were the intuitions upon which mathematical certainty could rest. We recognize here, of course, the concept of simple spatio-temporal continuity through which *psyche* was purged from *physis*, allowing the Cartesian compromise to be struck. In the philosophy of Kant, the compromise appeared reinforced in steel. Nevertheless, in Kant's own lifetime, a challenge was already being mounted.

When Kant proclaimed the inviolability of the spatiotemporal intuition, he was affirming not only the Cartesian idea of *physis* but also a much older system of measure basically concordant with Descartes's: the geometry of Euclid. In Euclidean geometry, the assumption of simple continuity is embedded at the deepest level (see Capek 1961, pp. 7–34). It was this geometric order that was being called into question at the time of Kant's death (1804).

From initial doubts raised about Euclid's parallel postulate arose alternative, *non-*Euclidean systems, such as the hyperbolic and elliptic geometries. During the nineteenth century, the Euclidean approach was further surpassed through the development of projective geometry, and beyond that, the highly general study of topology was inaugurated. Without getting into the complexities of the issue, it must be noted that these supersessions of Euclid's system were carried out in such a way that the continuity assumption was challenged only indirectly. Nevertheless, before the nineteenth century was completed, there were intimations that the role of discreteness could be played to the hilt, that *grossly* discontinuous structures were conceivable (such as the Cantor set, which possesses an infinity of singular points), structures referred to by some mathematicians as a "gallery of monsters" (see Mandelbrot 1977, p. 13).

The general point being illustrated is that the old intuition of simply continuous order was shaken substantially by decidedly *counter*intuitive notions emerging in the last century. In psychophysical terms, we witness again the upsurgence of the unextended *psyche*, but now see that behind the challenge of psi phenomena and that of modern physics was a direct challenge to the very structure of the human intellect. What has been the response?

In his own reflections on the matter, physicist David Bohm (1980) described the universally sanctioned response given when the order of thinking and perceiving that predated the Cartesian became suspect. The order of ancient Greece was predicated on the idea of the circle, voiced specifically in the conviction that celestial bodies trace perfectly circular orbits. "To be sure," noted Bohm:

> when more detailed observations were made on the planets, it was found that their orbits are not actually perfect circles, but this fact was accommodated within the prevailing notions of order by considering the orbits of planets as a super-position of *epicycles*, i.e., circles within circles. Thus one sees an example of the remarkable capacity for adaptation within a given notion of order, adaptation that enables one to go on perceiving and talking in terms of essentially fixed notions of this kind in spite of factual evidence that might at first sight seem to necessitate a thorough-going change in such notions. (Bohm 1980, p. 112)

I might add that in the notion of "epicycles" is a kind of mockery of the image of the circle, an extravagantly complex, wholly gratuitous replication of this image. Yet those bent on preserving cyclicity evidently were willing to pay the price.

In a similar manner, the idea of simple continuity has been gratuitously preserved in the present era. The Cartesian order would have disintegrated had discreteness genuinely been accepted into the cognitive/perceptual system. But could the discontinuity simply be ignored? The solution was to artificially append to the old intuition this counterintuitive intrusion, in much the same way the essential noncyclicity of heavenly motions was grafted onto the notion of cyclicity (through the artifice of epicycles), so as to preserve the latter. The desire to implement this accommodative "solution" can be seen as the motivating force that underlies the entire nineteenth- and twentieth-century mathematical program of formal axiomatics.

Kline (1980) identified the three main schools of mathematics that have adopted this approach: set theory, logicism, and David Hilbert's formalism. Their treatment of the concept of *actual infinity* well exemplifies their stratagem, for the idea that infinity can be realized (i.e., is not always beyond one's reach) certainly runs counter to Kantian intuition and, by the same token, is fundamentally expressive of discontinuity, as mathematician Georg Cantor implied in his assertion that the infinite can be obtained only by "stepping out of the series" (1976, p. 88). In founding set theory, by what method did Cantor contrive to "step out of the series," so as to attain the infinite set? By the method of axiomatic invocation. The actual infinite was merely defined into being by formal proclamation, that is, by fiat. Had Cantor done otherwise, had he sought an intuitively natural bridge between the finite series and the infinite, the simple continuity of the former would have been clouded by doubt, the old intuition threatened. *[Cantor's approach is discussed in detail in chapter 4.]* In like manner and for the same basic reason, Bertrand Russell axiomatically invoked the infinite in founding logicism, and David Hilbert followed suit in his effort to "purify" the work of his predecessors by "out-formalizing" the formalists.

The failure of the axiomatic method is well documented by Kline, whose assessment of logicism might well apply to *each* of the schools: "All in all, in view of the questionable axioms and the long, complicated development, the critics could with good reason say that logicism produces foregone conclusions from unwarranted assumptions" (1980, p. 229). Kline goes on to recount the telling criticisms of the axiomatic schools given by mathematical "intuitionists" like Brouwer and Weyl, the most fanatical of whom, for their part, were content to bury their heads in the sand and pretend that the Kantian dictum had never been challenged. These zealots rejected outright the concept of the actual infinite and, with it, dismissed many of the findings of modern mathematics without offering any new intuition that could fill the vacuum. Kline also details the devastating fate of Hilbert's formalist program at the hands of Kurt Gödel, the dashing of Hilbert's hopes for achieving self-consistency and completeness in mathematics by purely formal procedures.

But as I have already noted, it is formalism that continues as the accepted style of operating in mathematics and among theoretical physicists trained mathematically. We have found that in quantum physics, fundamental discreteness arises in the attempt to account for the behavior of physical systems. The officially prescribed formalism for treating the problem is a device known as "Hilbert space."

Consider a primary feature of the quantum domain: the nonlocalizability of elements, the intrinsic inability to determinately fix the position of a subatomic particle. Since the position of a system in an extended, infinitely differentiable continuum is always uniquely determinable, the basic quantum indeterminacy strongly suggests that microworld systems do not occupy distinct positions, that is, that microspace is not simply continuous. But discontinuity is precisely the quality that is anathema to Cartesian thinking. Therefore, a multiplicity of simply continuous "spaces" is axiomatically invoked to account for the "probable" positions of the particle—"it" is locally "here" with a certain probability, or "there" with another. Does this collection of spaces called "Hilbert space" truly preserve simple continuity?

Each subspace of the multispace expression is made simply continuous within itself to uphold the mutual exclusiveness of the alternative positions of the particle. Such subspaces must be disjoint with respect to each other, their unity being imposed externally, by fiat, rather than being of an internal, intuitively compelling order. Thus, in the name of maintaining mathematical continuity, a rather extravagantly *dis*continuous state of affairs is actually permitted in the standard formalism for quantum physics, an indefinitely large aggregate of essentially discrete, disunited spaces. Do not the self-continuous subspaces of the Hilbert formalism mock simple continuity in largely the same way the self-circular epicycles of antiquity mocked circularity? How explicit must the analogy be made before it is recognized that the ancient folly is being played out again in our modern context?

This is the epistemological crisis we confront. It arises in essence from our persistent effort to deal with non-Cartesian phenomena in the old Cartesian way. The crisis in *psi modeling* is closely linked to that in modern physics inasmuch as the most ambitious and interesting models of psi (e.g., Walker 1973;

Mattuck 1977), however they may differ in their details, are generated in the same formalistic style as employed in physics proper (some making direct and explicit use of the Hilbert formalism). What we have seen is that this style of modeling is by no means philosophically neutral but can be viewed as a rearguard attempt to save the Cartesian compromise in the face of circumstances that cry for its abandonment. Operating under the still powerful notion that "mind is mind, body is body, and never the twain shall meet," the formalist just cannot accept unextended *psyche* in the midst of *physis*. His/her basic motive is to express the unextended in terms of the extended, that is, to reduce *psyche* to *physis*, not to reconcile them genuinely. At bottom, a classically Western monism not a harmonizing dualism, is the formalist's goal. In the failure to achieve this goal, in the inability to extrapolate classical monism into the arena of modern realities, a dualism does arise—in fact, an infinite pluralism. But it is one of deficiency, more a *mockery of monism* than anything else, one that seems to be saying to us, "this is the price we must pay for our continued denial of *psyche*." It is my conviction that the price is too high.

META-MODELING AND THE PROSPECT FOR "EPISTEMOTHERAPY"

In sum, our epistemological crisis is a crisis concerning the *psyche*. Several hundred years ago, *psyche* was relegated to obscurity so we could concentrate our efforts on reinforcing the extended, physical aspect of our being. Since then, our activities, operations, transformations of our environment and each other have been predicated on the underlying assumption that physicality as such will be preserved, remain untransformed. The deeper our preoccupation with *physis*, the more unthinkable the *psyche* became, the greater our predilection to reduce it to extended expression, or simply deny its existence. But this is precisely what we no longer can do, if my interpretation of the anomalies currently besetting us is correct. However deeply ingrained our commitment to the Cartesian compromise and its *psyche*-denying consequences, it appears there is no choice now but to disengage ourselves from that old allegiance and *accept* the *psyche*.

There is a choice to be made though, a critical one involving our manner of acceptance. We might view the upsurgence of *psyche* as an indication that all is lost, that we must surrender reason and objectivity, dissolve into chaos. The manifestation of *psyche* in our midst might be seen as the signal to forsake our discriminative powers, as license to engage in injudicious and arbitrary acts, to indulge ourselves in the irrational. This would be mere capitulation, of course; here *psyche* would be accepted in a purely passive fashion.

The alternative is to accept *psyche* actively. While no attempt would be made to deny the collapse of the old intuitive base, neither would we simply allow ourselves to devolve, to lapse into the "nonintuitive," for the burgeoning of *psyche* would be seen to reveal our potentiality for creating a *new* intuition: it would be taken as an invitation to evolve. Put differently, rather than clinging to our old physical identity in the vain attempt to bar *psyche* (as we do in perseverating upon the classical epistemology), or simply abandoning *physis* to *psyche*, by active acceptance we would be permitting a creative interplay of *physis* and *psyche* through which a fresh identity could emerge.

I am suggesting a program of "epistemotherapy"[1] that can be regarded as an extension of the approach to individual psychotherapy advocated by C. G. Jung. For Jung, the ultimate goal of psychotherapy is to achieve a balance of ego (associated with *physis*) and Self (associated with *psyche*). A fundamental method recommended is that of active imagination (Jung 1969). Deliberately and without just dissolving, the ego becomes receptive to what lies beyond it. It does this by allowing aspects of the Self to take concrete form as psychically potent, nonreductive images or symbols, representations of the Self that nevertheless are not mere ego contrivances. By thus permitting positive expression to the Self—an uncontrived, intuitively coherent embodiment— the presence of the "unconscious" is authentically felt within conscious awareness and a therapeutic centering effect is realized.

But to address effectively the problems discussed in this paper, the symbolic realization of *psyche* within *physis* currently required could not be enacted merely within the limited sphere of an individual ego's private way of knowing (as in the conventional practice of Jungian psychotherapy); it would need to be extended to the domain of the collectively shared, public order of knowledge, to be brought to bear at the foundations of science and mathematics. In so doing, individual depth psychotherapy would become "epistemotherapy."

The psychologist Marie-Louise von Franz was proposing such an extension, I believe, when she advised that Jungian analysts become students of physics and that physicists adopt a Jungian strategy (1975, p. 45). Parapsychologist Rhea White (1960) made a similar proposal for the practitioners in her own field of study: to approach psi from the "inside out," that is, from within the *psyche*. Were we to heed these pleas, we might surmount Cartesian epistemology and engage in what elsewhere I have called "meta-modeling" (Rosen 1981).

We know that Cartesian modeling is undertaken for the purpose of relatively short-term analysis, prediction, and control, activities wholly appropriate when operating within an established intuitive context. Here models are constructed to facilitate manipulations, with the manipulator's identity being preserved. As long as the intuitive framework holds together, retains its integrity, the identity of the modeler is secure. Working from within an absolute intuition of *physis* (expressed through self-evident axioms of the Kantian variety), modelers can detach themselves from the relativity of events transpiring in *physis*. By this act of abstraction, they gain free reign to devise whatever models of such dynamics are useful in their identity-maintaining maneuvers.

It is the breakdown of the given intuitive context that furnishes the impetus for meta-modeling. While identity is absolute and process relative in the realm of the fixed intuition, the converse becomes true in transcending that realm. A nonrelative meta-process operates to transmute identity. Therefore, for the prospective modeler of this meta-intuitive process, the attitude of detached manipulator is hardly appropriate. In this case, modeling does not mean contriving with a free hand in the interest of preserving identity but entails a surrendering of identity that is nonetheless disciplined, one that avoids the mere dissolution of identity, fosters the re-creation of identity by properly embodying the meta-process. The implication of "proper embodiment" is that models or symbols exist that are neither arbitrary conventions (as in logico-formalism) nor self-evidently intuitive in the Kantian sense of the extensionally actualized, the already-formed and immutable. These "meta-symbols" derive their intuitive validity from the deeper source, the intensional domain of potentiality, from the flux of meta-process that Jung would call *Self*. Such symbols are termed *archetypes* in Jungian parlance, and their function is to transform the symbolizer therapeutically.

In traditional depth psychotherapy, active imagination may be beneficially implemented through a model, drawing, or painting individually fashioned by the participant. But for the constructive reordering of our collective mode of knowing envisioned in epistemotherapy, the most universal form of archetype would seem to be required. Indeed, toward the end of his life, Jung himself apparently saw the need for discovering more universal archetypal forms. And his colleague, von Franz, carried this work forward after his death, her explorations culminating in the book *Number and Time* (1974). Number is interpreted here in a non-Western, qualitative manner and hypothesized as the primordial organizing principle for the *unus mundus* or "hidden continuum" from which less fundamental archetypes emerge.

But can any ordinary number effectively represent the profoundly paradoxical *unus mundus?* About fifteen years ago, mathematician Charles Muses (1977) introduced the notion of "hypernumber" (the imaginary number i being his point of departure). Related to Muses's idea, I believe, is the more recent attempt of Frescura and Hiley (1980) to mathematize David Bohm's "enfolded" or "implicate order." My own work on the infinite *[see chapter 4]* might also be relevant. As I view them, such efforts appear archetypal and could well prove epistemo-therapeutic in their intimation of the mathematical content for collective meta-modeling. These radically non-Cartesian concepts surpass the notion of simple continuity and, in so transcending, make room for an authentic acceptance of *psyche.* By the same token, when we examine them closely, we realize that they offer a natural answer to the question about "unified fields" in theoretical physics *[see chapter 5]*, for they address themselves meaningfully, without artifice, to the greatest obstacle to unifi-cation, the problem of discreteness, of the irreducibly infinite—which is one and the same as the problem of *psyche*, as I have tried to show. In sum, where both logico-formalism and classical intuitionism have failed to lead us out of our Cartesian cul-de-sac, the emerging meta-intuitive/archetypal approach may succeed.

10

Parapsychology's "Four Cultures": Can the Schism Be Mended? (1984)

INTRODUCTION

In his celebrated essay *The Two Cultures*, C. P. Snow (1964) commented on the deep division between the world of science and the literary or artistic world. Generally speaking, the scientific outlook is externally, objectively oriented, linear, geared to collective advance, optimistic about the progress of humanity. By contrast, the artist looks inward to a private reality. Largely unconcerned with the march of progress, s/he responds to the call of the ineffable, seeks to give voice to the unvoicable, and in the process may travel convoluted and idiosyncratic paths. Snow viewed this split between mind sets as a tear in the fabric of society bearing grave consequences; his hope was that somehow the schism could be mended, our social schizophrenia healed.

Now, it appears that parapsychology has its own "cultural schism," a division reflecting in microcosm the broader bifurcation examined by Snow. This split within the parapsychological community was brought out by Stanley Krippner in his 1983 presidential address to the Parapsychological Association (Krippner 1984). Adopting Mitroff and Kilman's (1978) Jungian typology for social scientists, Krippner identified *four* approaches to research in parapsychology: "The sensing-thinking type is the Analytical Scientist; the intuition-thinking type is the Conceptual Theorist. The Conceptual Humanist is the intuition-feeling type and the Particular Humanist is the sensing-feeling type" (p. 155). In terms more applicable to society at large, perhaps we could speak of the Scientist, the Philosopher, the Artist, and the Social Activist, respectively. Then employment of the Jungian scheme of classification would enable us to understand that our "cultural schism," being fourfold in character, is actually more complex than Snow estimated.

By calling attention to the diversity of operating styles within parapsychology, Krippner hoped to help us come to terms with the unhealthy tendency for one style to dominate to the detriment

167

and devaluation of others. Moreover, Krippner advocated a systems approach in which problems of psi would be addressed by teams of researchers whose distinctive and complementary styles would allow a measure of success unattainable by any one style alone. In the strategy envisioned, the differences among the styles would be preserved, the Analytical Scientist operating exclusively in the sensing-thinking mode, the Conceptual Theorist in the intuition-thinking mode, and so forth. Yet from the Jungian perspective, the four modes of relating stand in fundamental opposition; using Krippner's own language, they are "antithetical psychological processes" (Krippner 1984, p. 153). Under ordinary circumstances, the antipathy between opposing functions (thinking vs. feeling, sensing vs. intuition) is so great that coexistence is not possible; in the psychodynamics of the individual, one member of a dipolar pair tends to dominate, the other being denied by consignment to the unconscious. What does this auger for the *group* dynamics of a team of psi researchers composed of antithetical types? Is it realistic to expect that each member, working in a style independent from and unreconciled with the styles of the others, can make a constructive contribution to the overall effort of the group? Can we really have a cohesively func-tioning group when typological relations among members are essentially external? Put differently, if the parapsychological enterprise is a dialectical process of growth and development (as Krippner appears to suggest [1984, p. 162]) can it be limited to a mere juxtaposition of antithetical elements, or must it not advance to a *synthesis* of these opposites, if maturity is to be attained?

When the Jungian strategy is pursued beyond the simple identification of polarized types, as Rhea White (1985) has done, it is clear that the direction of development foreseen is toward synthesis, an integration of antithetical functions through the activation of a "fifth," *transcendent* function (Jung 1969). Jung referred to this growth process as "individuation," viewing it as the primary path to full and authentic maturity. This is the process I would like to explore and attempt to clarify in my presentation. My main thesis is that individuation, properly understood, is precisely what is needed for mending the schism among society's "four cultures," and that parapsychology— despite its tendency to continue operating in the customary, unindividuated manner—is particularly suited to provide creative leadership in this endeavor because, both that which parapsychologists seek to know (their subject-matter) and the required mode of parapsychological knowing, directly entail the "transcendent." At the conclusion of my paper, I will address myself to the question of individual vs. communal transformation and, in that context, to the mistaken idea that individuation exclusively concerns the former.

TRIADIC ORTHOGENESIS AND INDIVIDUATION

Before going any further, it might be advisable for me to identify my epistemological grounding in explicit terms, so as to avoid being misunderstood. I would maintain that while the currently favored epistemology of logico-empiricism is ill-equipped to address meaningfully the question of developmental process (since, in this approach, process is treated essentially as deriving from fixed product, rather than being accepted and dealt with on its own terms [see chapter 9]), there is another tradition now winning wider recognition that may meet with more success. I am speaking of the "evolutionary epistemology" exemplified in the writings of Henri Bergson (e.g., 1966), Milic Capek (1961), and, notably, in the work of Alfred North Whitehead (1978), who abandoned analytical philosophy for what he called "speculative philosophy," after becoming disenchanted with his own ambitious effort (in Whitehead and Russell 1925–27) to found mathematics on a sure logical footing (see the critique of logicism by the mathematician Morris Kline [1980, pp. 216–30]). (Recent statements of evolutionary epistemology that cogently assess logico-empiricism include Hooker 1982, and Bjelland 1986.)

In the evolutionary epistemology of consciousness researcher Ken Wilber (1977, 1980, 1981), an attempt is made to clarify developmental process by demonstrating its triadic character: "The overall sequence of development [is] from nature to humanity to divinity, from subconscious to selfconscious to superconscious, from prepersonal to personal to transpersonal" (1980, p. 52). Functionally, we may speak of a pattern of movement from lack of differentiation, to differentiation, to integration. The initial state is one of primordial wholeness, an unfocused, infantile condition devoid of an awareness of boundaries or distinctions. Transition to differentiated functioning is a step "up from Eden," as Wilber (1981) would say; the distinctions that now can be drawn are necessary for sustaining life and enriching its quality. But the "timing mechanism" of evolution is such that the differentiative mode serving to enhance life eventually calcifies into a destructive parody of itself. Life-affirming differentiation becomes fragmentation, and this is a sign that the time has come to enter the third phase of orthogenesis, that of integration. Here, the reflexive character of the developmental pattern becomes evident, for integration entails a recovery of wholeness. However, this is not to be viewed as a simple regression, a return to "Eden," a recapturing of lost innocence. Rather, by entering the integrative phase, we "go back" to wholeness by "going forward," that is, by bringing antecedent phases into *harmony* so that the duality of the undifferentiated and differentiated is realized nondually *[the notion of "nondual duality" is the central theme of chapter 11; see also chapter 14]*. Therefore, while wholeness would be achieved through integration, there would be no lapse into oblivion; the clarity gained from differentiation would not be forfeited.

The general idea of triadic orthogenesis is reflected in the work of numerous thinkers. It was the basis of Herbert Spencer's interpretation of Darwinian evolution, and it has been used by many students of the growth process, with psychologist Heinz Werner (1948) enunciating it perhaps most explicitly. Ken Wilber, from his own triadic perspective, examines the "pre/trans fallacy," a widespread tendency among theorists "to confuse prepersonal and transpersonal dimensions" (1980, p. 53). It is on this ground that the Jungian concept of development is critically evaluated. According to Wilber (1980, p. 56), Jung consistently mistook the notion of transpersonal awareness for a purely prepersonal mode of being. Wilber believes that as a consequence, Jungians often are predisposed to portraying individuation in naively idealistic terms, as a simple dissolving of all distinctions into an utterly featureless totality. However, although Wilber essentially is attributing to Jung the recognition of only two domains, the differentiated ego and the undifferentiated Self, Jung actually was asserting that the goal of development is not to leave ego behind for pristine Self, but to achieve a *balance* of these; here there is at least an incipient recognition of a third, integrative mode of operation. In fact, turning the tables on Wilber, it could be argued that his misassessment of Jung derives from his own failure to understand the reflexive character of development, that is, that the third, transpersonal "realm," while transcending the prepersonal and personal states, is not simply distinct from these but constitutes their reconciliation, a grasping of their nonduality without confusing them. A further indication that Jungians indeed can appreciate the triadic nature of development is found in depth psychologist Marie-Louise von Franz's (1975) sharp insight into the order of differentiation she criticizes: "The rationalism of the 17th century. . .had one advantage after all: it drove father-spirit and mother-matter so far apart that now we can reunite them in a *cleaner* way" (p. 42; emphasis added). As I interpret it, such a "clean reunion" of the undifferentiated and differentiated is just what is entailed in individuation, seen as the integrative culmination of triadic orthogenesis.

[In more recent years, debate on the nature of transpersonal development has been joined by philosopher Michael Washburn (1988), who has systematically elaborated an explicit alternative to Wilber's approach.]

A "GENETIC CODE" FOR OUR CONTEMPORARY SCHISM

The cultural schism described by C. P. Snow may be understood in the developmental context proposed. Our current condition of fragmentation—within parapsychology, in the general scientific community, in society as a whole—may be interpreted as a symptom that we are approaching the limits of the differentiative mode of functioning and that it may now be necessary for us to enter the phase of integration, achieve individuation. But in meeting this formidable challenge, it seems we must grasp the fact that the fragmentation we face is a "fragmentation squared," as it were. That is to say, the schism that needs to be mended is no simple division of a whole into parts but, rather, a *higher-order* division in which the "parts" themselves are dividedness and wholeness. I propose and will presently demonstrate that these "parts," elucidated in a Jungian context, constitute the basic terms of a "genetic code," so to speak, from which we may read Jung's preindividuated "phenotypology."

To begin, let us consider Jung's grouping of the four psychological functions into "nonrational" and "rational" categories. In broad phenomenological terms, we may view the nonrational functions of sensing and intuiting as *presentational*, since they entail an immediate reaching out to, grasping and taking in of that which is other in the field of experience. By contrast, the rational functions of thinking and feeling are *re-presentational*, involving a reflection on, and (e)valuation of what is given. Or perhaps more simply and basically, we could say that the non-rational operation is a knowing *of*, while the rational is a knowing *from*, that is, from a given perspective or evaluative base. Now, on the nonrational side, it is evident that to sense the world is to experience it discretely, to divide it into segregated units with sharply drawn boundaries, while intuiting the world means experiencing it as an undivided whole. Regarding the rational functions, the discrete, separative, "partitive part" is feeling; here experience is assessed subjectively, valued from the idiographic perspective of this particular body as distinct from others. Antithetically, the evaluative ground of thinking is objective: Judgments are made, not from the concrete vantage point of the individual as such, but from the frame of reference of an abstract, disembodied community of knowers arriving at conclusions about the known through intersubjective agreement.

We may term the domain of the divided or diverse, *D*, and that of the undivided or unitive, *U*. In table 10.1, these terms are used to code the "four cultures" as expressed in Jung's typology and in the Mitroff and Kilman (1978) adaptation of Jung employed by Krippner (1984). The first column of table 10.1 gives the Mitroff and Kilman social science types (Analytical Scientist, Particular Humanist, Conceptual Theorist, Conceptual Humanist), with the Jungian prototype parenthetically indicated for each; columns two and three provide the *D/U* coding for corresponding nonrational and rational functions, respectively. From the table, we see that for both the Analytical Scientist and the Particular Humanist, the world is taken in through the senses and, therefore, known concretely in its particulate diversity, whereas the non-rational predilection of the Conceptual Theorist and Conceptual Humanist is holistic. But the Scientist and Theorist reflect on their data "objectively," that is, through an abstract, intersubjective unity of knowers, whereas the judgmental base of the Conceptual and Particular Humanists is *subjective*, concretely and individualistically embodied.

By the insight it gives into the deep structure of our phenotypic, psychocultural division, the *D/U* "genetic code" should shed light not only on the nature of the current schism but also on what may be entailed in its mending. However, before we proceed to the question of schism-mending individuation, I

Table 10.1

"Genetic Code" for Mitroff/Kilman Jungian typology

Type	Nonrational Function	Rational Function
Analytical Scientist (sensing-thinking)	D	U
Particular Humanist (sensing-feeling)	D	D
Conceptual Theorist (intuiting-thinking)	U	U
Conceptual Humanist (intuiting-feeling)	U	D

Note: U = Unity, D = Diversity

would like to re-emphasize that from the standpoint of developmental pragmatism, the double polarity described is not in itself to be negatively judged—it is our evolutionary *timing* that is problematic. To use the Wilberian metaphor once again, the polarization of human functioning was initially a step "*up* from Eden." A useful purpose was served by overcoming the primal confusion of sensing and intuiting (i.e., of presentational diversity and unity), and of feeling and thinking (re-presentational diversity and unity). Yet, in the presentational realm, the sense-based divisions once so helpful to us in managing our world have hardened into the fragmentation that now seems to rob our world of meaning. By virtue of the same extreme polarization of functioning, the wholeness of intuition that gave fresh perspective to life has become an obscureness, an opaqueness, a mystery that no longer can be penetrated. In the sphere of re-presentation, a healthy individuality of feeling has turned schizophrenically idiosyncratic, while the process of consensual agreement and validation based on thinking has rigidified into conventionalistic dogma. What all this appears to indicate is the need to advance to the third, integrative phase of orthogenesis. It seems we must surmount the higher-order division of the divided (D) and undivided (U), achieve higher-order wholeness through a schism-healing act of individuation. [*The concept of 'higher-order wholeness' is the primary concern of chapter 14.*]

THE PHENOMENOLOGY OF INDIVIDUATION

For both presentational and representational functions, individuation would entail a fusion without *confusion* of the *D* and *U* modes. The sharply drawn, strict dualism of *D* and *U* now existing would become a *nondual* duality. To convey a graphic, experientially immediate understanding of what this means, I turn to the medium of visual phenomenology. . .

[In the ensuing passages of the original paper, the exercise of integrating the perspectives of the Necker cube is performed, as in chapters 5 and 6. Presently, it is seen as a way of symbolically fleshing out the higher-order unity of diversity and unity proposed to be the essence of individuation.]

So the proposition I am offering is that the experience of "one-sidedness" produced with the Necker cube constitutes an immediate grasping of the relation of "nondual duality" that would be entailed in the actual reconciliation of *D* and *U*. I suggest that viewing the cube in the manner indicated gives us a direct glimpse of the specific nature of higher-order wholeness, the "shape" of the individuation required for healing our fourfold psychocultural schism (the double dipolarity). By so witnessing the depth of interpenetration of *D* and *U* opposites that nevertheless retain their distinctness—the "clean reunion" of "mother-matter" and "father-spirit"—a structural/phenomenological clarification is gained of the climactic phase of orthogenesis, which, until now, has been described only in general terms. And should not such a clarification of the phase of integration itself be an integral part of the process? The special significance of parapsychology for this process will now be discussed.

INDIVIDUATION AND PARAPSYCHOLOGY

How may we interpret integration of the nonrational functions of sensing and intuiting? What would a "nondual duality" of sensation and intuition involve? Evidently, it would mean that in a single act of apprehending the world, we would be both dividing it and experiencing it unitively, encountering its outer, (di)visible face, and its inner, in(di)visible or holistic face; outside and inside would be grasped as but a single side. By virtue of the outer (i.e., sensory) aspect, such an experience would have "veridical substance"; that is, it would possess the quality of ingesting "hard data," information which is immediately factual or actual, experienced as objectively given. But at the same time, integrative grasping would constitute—to quote Jolande Jacobi's description of intuition—"an 'inner perception' of the inherent potentiality of things" (1962, p. 12). Does there currently exist such an "inside-out," unitary mode of experiencing, one that bridges the gap between fact and possibility, the actual and the potential? I suggest that there does and that it is a form of functioning centrally relevant to parapsychology, a form of *psi* functioning. In the present context, perhaps the most obvious example is precognitive extrasensory perception. Unlike sensory perception which is limited to immediate actuality, precognition does involve a direct apprehension (as opposed to a predictive extrapolation from thinking) of the "inherent potentiality of things." Yet precognition is no mere act of imagination, not just a hunch about the future (Jung associated intuition with hunches [1964, p. 61]). *Like* sensory perception, precognitive viewing is veridical.

What of the integration of the *rational* functions of thinking and feeling? If integrative presentation means a simultaneous knowing of the divided and undivided, integrative *re*presentation would mean a simultaneous knowing *from* divided and undivided *perspectives*. I propose we call such knowing "transsubjective" (using Wilber's [1980] term). Operating *sub*jectively (through feeling), perspective is embodied in the concrete particularity of the given individual to the exclusion of others, whereas *inter*subjective functioning (thinking) unifies perspectives by abstraction or disembodiment (as I observed earlier). Note that the unity achieved in the latter is at the *expense* of concrete individuality; therefore, the intersubjective mode does not constitute a direct linking of individuals. Rather, the individuals who make up the intersubjective community share their common perspective indirectly, through the mediation of symbols that must be encoded and transmitted from one individual to another. It is by means of symbolic abstraction that individuals are intersubjectively linked; in their concreteness, they remain diverse. However, individuals operating from a *trans*subjective base would be joined (the *U*, or thinking aspect) directly from the core of their concreteness (the *D*, or feeling aspect), giving a subject-to-subject interpenetration that would require no encoding, transmission, or decoding for perspectives to be shared. The transsubjective community would be diversely embodied in its very communality

to produce the nondual duality of diversity and unity characteristic of individuation.

As we did for the case of the nonrational functions, we may
ask if there currently exists a rational form of integration such
as that just described. Again, a form of psi appears to suggest
itself. Jung's distinction between nonrational and rational operations seems paralleled by the customary distinction parapsychologists have drawn between "object-to-mind" and so-called "mind-
to-mind" modes of psi, respectively. The former is generally
termed "clairvoyance," one type of which would be the precognitive
clairvoyance referred to above; the latter is designated as
"telepathy." In contrast to clairvoyant functioning, the "mind-to-
mind" or "subject-to-subject" telepathic mode of awareness
would not entail a paranormal knowing *of* that which is other,
as much as a knowing *with* other, a knowing from another's
perspective; therefore, in the Jungian sense, it would be "rational."
And clearly, this telepathic unification of perspectives would be
no mere act of abstraction linking individuals indirectly, as in
ordinary thinking; rather, it would possess that very quality of
concretely direct, core-to-core connectedness attributed to rational
integration.

Of course, parapsychologists have not found it easy to sharply
distinguish telepathy from clairvoyance. In fact, the categorical
delineation of *any* form of psi, including psychokinesis, has
proven elusive to us. Yet I submit that if psi *is* intimately related
to individuation, this is just what we should expect, for the
wholeness achieved through individuation indeed would be of a
radical order. The unity attained would not be limited to an
integration of functions *within* rational and nonrational categories
but would occur *between* them as well. Whereas, in the differentiative phase of orthogenesis, a strict trichotomy of presentation,
representation, and action is possible, with integration, these
three modalities would be unified in the nondually dual manner
discussed—woven together as distinct aspects of one and the
same reality. The scope of this essay does not permit me to
elaborate further on this. Unfortunately, having just scratched the
surface of psi's specific relation to individuation and of the
thoroughly holistic character of individuation itself, I must begin
my concluding remarks.

Within the developmental context I have suggested, psi
generally would be understood as a precursor of the third, inte-

grative phase of orthogenesis. This would be consistent with the transient nature of psi, its phantomlike undependability. Psi is ephemeral because it is *embryonic*. Naturally, the "embryo" of higher-order wholeness, of individuation, cannot be brought to term while we are still in the grip of fragmentation. And all of us, in one way or another, to one extent or another, remain in that grip, continuing our resistance to the thoroughgoing transformation that is indicated. Students and researchers of parapsychology surely are no exceptions. But in my view, what *is* exceptional about parapsychology is that the phenomena of this field *demand* our transformation, if we are to do justice to them.

As matters now stand, both in our interpretation of our subject matter and in the methods of study we adopt, the pre-individuated mind frame persists, with its splitting of unity and diversity. Consequently, regarding our understanding of psi phenomena, we are limited to the four philosophical alternatives recently summarized by John Beloff (1985). We view psi as belonging entirely to the (di)visible realm, as in materialism or physicalism, or to the in(di)visible realm, as in idealism, or psi is presumed to be the product of an interaction between these domains, said realms being seen as strictly dichotomous, as in Cartesian dualism. Elsewhere (Rosen 1985), I have demonstrated that none of these alternatives provide an adequate basis for comprehending psi, that an integrative, nondually dual approach is required. *[In the next chapter, Beloff's four alternatives are examined more closely.]*

As concerns our methodology, the perspectival base from which parapsychologists evaluate their subject matter, once more the *D-U* schism is in evidence. Thus, the only research perspectives we recognize are the intersubjective (the unitive way of the Analytical Scientist and Conceptual Theorist), or the subjective (the diversified approach of the Conceptual and Particular Humanists), or some loosely complementary admixture of these that would preserve the simple distinction between them. Restricted as presently we are to these antithetical alternatives, the telepathic mode is considered merely as a phenomenon for us to study, rather than a new *methodological* possibility, a potential *base of operations* that would constitute a transsubjective synthesis of the older strategies, which, by themselves or in weak combination, are not equal to the challenge of psi. I hasten to admit that adopting a transsubjective methodology is easier said than done, but perhaps the first step toward doing it is saying it, acknowledging the genuine need for it.

It is my conviction that if the challenge of psi can be met, then, as Rhea White (1983) envisioned in her glimpse into the future of parapsychology, we should emerge as leaders, facilitators of creative change, in fact, as healers, since the change of which I speak is a realization of wholeness. What I am proposing is that progress in parapsychology hinges on a fundamental shift from being an enterprise whose first order of business is the accumulation of knowledge (to be *followed* by applications) to being one in which the *primary* goal is a therapeutic mending of our deeply destructive polarization.

Since the program of therapy I am advancing has taken its cue from Jungian *psycho*therapy, I will end by explicitly addressing the question implied at the beginning of my presentation. Are we not to think of psychotherapy as chiefly a matter of *individual* growth and development? Is not the proper province of psychotherapy the life and problems of the particular person? We are now in a position to see that while the conventional view of and approach to psychotherapy does have this narrow focus, the restriction is itself expressive of the differentiative, pre-individuated way, presupposing as it does a categorical separation of individual and communal levels, of *D* and *U*. Transcendent, individuated functioning, seen as the ultimate aim of psychotherapy, by its very nature would entail a healing of this split. Such a psychotherapy would be inherently psycho*social*, for it would result in an "individualization of the community" and a "communalization of the individual." That is to say, an embodied intimacy would replace the faceless uniformity now prevailing in community, and the schizophrenic alienation of the person would be supplanted by a transpersonal, core-to-core union with other.

11

Psi and the Principle of Nondual Duality (1987)

INTRODUCTION

Since the time of its inception a little over a hundred years ago, modern parapsychology has been seeking to establish itself as a legitimate field of scientific inquiry. But despite a few token indications of acceptance (such as the somewhat grudging admission of the Parapsychological Association into the American Association for the Advancement of Science in 1969), not much tangible progress appears to have been made. A negatively skeptical viewpoint has prevailed, with parapsychology remaining on the fringes of organized science. Yet it well may be asked whether parapsychologists realistically can expect to gain a more secure position when they themselves frequently have been as skeptical as their most vocal critics.

The tendency among parapsychologists to deny their own phenomena is well known and widely acknowledged. Parapsychologist D. Scott Rogo (1977) has discussed the "will to disbelieve" or "morning-after syndrome" in which findings become clouded by doubt, despite a strong conviction of authenticity at the time of observation. The general subject has been treated by White (1985), Eisenbud (1967), LeShan (1984), Rosen (1979), and a number of others. At the 1984 Annual Convention of the Parapsychological Association, a roundtable was devoted to this issue. There, as one of the participants, I emphasized that the problem cannot effectively be addressed in a vacuum, that is, without a genuine understanding of its basis and origin. I proposed that some "archeology" may be in order, some digging into the source of parapsychology's self-denial.

In the following presentation, I will attempt to begin this "archeological expedition" in earnest. My basic thesis is that psi phenomena are fundamentally incompatible with the general philosophico-religious orientation that has pervaded Western thinking for centuries, an outlook that has had considerable influence even in the East. After examining the paradoxical character of psychic experience when viewed from within the predominant mind frame, I shall explore an alternative framework that promises to be more harmonious with psi.

But before proceeding, I take note of the fact that the approach I am planning to call into question prescribes the basic method by which all questioning is supposed to be done. In the interest of gaining positive knowledge, achieving the highest possible level of certainty, one must strive for maximum clarity, be explicit, give sharp definitions in advance, and, in more formal inquiry, set forth well delineated propositions from which theorems are generated and put to empirical test. Now I would argue that the striving for this sort of certainty, in large part, is an act of compensation for the sacrifice of a deeper certainty. Underlying the method of questioning, of knowing, sanctioned for the entire enterprise of philosophy, science, and technology as practiced in our culture from the earliest times is the *un*questioned presupposition that the knower essentially is *separate* from that which is known. In this loss of internal relatedness, of intimacy with the known, knowing is relativized, creating the profound sense of insecurity we have been doing our best to overcome by the forging of tight correspondences. Yet however tight these logico-empirical linkages may be, in essence they are *external* linkages.

So to question the traditional approach to knowing at the most fundamental level is to question the assumption that the knower is separate from the known. Perhaps, by identifying this basic presupposition undergirding our tradition in the introductory portion of my presentation, I am getting a little ahead of myself, but I do not wish to be misunderstood by those who might be expecting me to build my entire case on a set of explicit definitions or discursive logic or citations of large numbers of empirical studies. It is this approach to gaining certainty that I am *calling into question* so as to pave the way for exploring the possibility of regaining the more intimate way of knowing. Indeed, I intend to demonstrate that psychic knowing constitutes just such a mode of intimate awareness. Accordingly, the approach that I will take will have a *non*discursive aspect. Matters will be left somewhat implicit, as they necessarily would be in more artistic forms of expression. As philosopher/physicist David Bohm (1980) might say, my mode of operation will have an *implicate* quality, for to explicate is, by itself, simply to separate.

I believe we who are interested in psychic functioning understand fairly well that field and laboratory research in psi is a participatory affair. Parapsychologists such as Rhea White (e.g., 1976), in drawing attention to the likelihood that boundaries between and among psi subjects and experimenters are quite arbitrary and artificial, have pointed to the need for researchers to implicate themselves consciously in psi, attempt to *enter into* psi rather than pretend merely to be studying it from an external perspective. Less clear to us, perhaps, is that an implicate approach also may be required in *philosophical* psi research and this would amount to a style of doing philosophy in which the philosopher cannot be restricted to the axiomatic, discursive, logico-empirical method. For that is the method that imposes beforehand a categorical schism between researcher and researched, knower and known.

THE CARTESIAN TRADITION AND THE PARADOX OF PSI

While the philosophy of schism long predates the period of the Renaissance in its basic thrust, it was powerfully reinforced by developments occurring at that particular juncture in the history of Western culture. A central figure to emerge within the philosophical tradition of the Renaissance, an individual whose thinking "left its mark on the whole subsequent history of philosophy" (Jones 1952, p. 686), was the French mathematician René Descartes. It is the Cartesian manner of formulating our schismatic condition that is most familiar to us today and probably has had the greatest impact. The knower is identified with the inner reality of the self, with *psyche* or mind, while the known is associated with that which is other, with the external world of the physical, the bodily. And between these domains lies a gulf that cannot be bridged. . .

[*At this point in the original essay, Cartesian dualism and the paradoxical, non-Cartesian character of psi are elaborated in much the same manner as in chapter 9.*]

In sum, the psi phenomenon is essentially non-Cartesian in nature. On the one hand, if we assume its validity we must admit that it is directly implicated in physical reality and thus [*unlike fantasies or dreams*] it cannot be relegated to a domain of pure mentation utterly divorced from space and time. On the other hand, psi is not *contained* by space and time, not completely limited by the constraints of the extensive physical continuum. Its significant deviations from the proscriptions of *physis* warrant that we ascribe to it a *non*physical aspect *[see chapter 9]*. Evidently then, we are obliged to think of the psi phenomenon as a *hybrid* occurrence owned neither by *physis* nor *psyche* alone, engaging both of these domains, bridging them in a manner quite unfathomable to the purely separative Cartesian point of view.

Having described the Cartesian position and indicated its profound influence on our thinking, I now want to reaffirm that the philosophy of schism is not peculiar to Descartes. Though the dualistic orientation may most readily be recognized in his specific writings, in fact, there is an important sense in which it is implicit in *all* major schools of modern philosophy, doctrines that superficially may seem to compete with the Cartesian. Evidence for this deep-lying, far-reaching dualism also can be found in equivalent approaches to *theology.* Moreover, as I noted above, the philosophy of schism predates the modern era, being traceable to the very origin of philosophical thought in both the West and the East.

THE PERVASIVENESS OF THE DUALISTIC ORIENTATION IN PHILOSOPHY AND RELIGION

In a recent paper, "Parapsychology and Radical Dualism," parapsychologist John Beloff (1985) attempted to support a Cartesian interpretation of psychic functioning. In replying to Beloff (Rosen 1985), I called attention to his unexamined and erroneous presupposition that psi is exclusively a phenomenon of mind, understood as categorically distinct from body. Then as now, I argued for a *hybrid* interpretation of psi that would require us to overcome the powerful influence of schismatic thought. Of immediate interest at present is Beloff's statement of the philosophical approaches that are possible with regard to psi, a set of alternatives he apparently viewed as exhaustive and mutually exclusive.

The forced choice Beloff gave us is among pure materialism, idealism, physicalism, and radical dualism. It is customary to consider the first two approaches to be forms of monism, the latter two as forms of dualism. The members of the monistic pair are simply and diametrically opposed. The materialist categorically affirms physical reality while denying *psyche*, whereas the idealist adopts the converse stance. Unlike the materialist, the physicalist does posit a distinct mental domain but denies autonomy to it, viewing its manifestations as mere epiphenomena of the physical. And of course, radical dualism is the Cartesian position in which *physis* and *psyche* are ontologically equal but utterly separate spheres of operation nevertheless assumed able to interact. Note that the physicalistic type of dualism is less radical than the Cartesian only in the negative sense of supposedly depriving the *psyche* of independence, not in the sense of overcoming the sheer separation of mind and body through a positive insight into their reconciliation.

The fact is that despite their diversity of content, *none* of the four alternatives set forth by Beloff overcome the *formal* separation of mind and body. In each there is the same tacit understanding of the form in which both mind and body are to be construed: They are to be predicated as simple, mutually exclusive categories of being. Although the monist predicates one category by negation, that is, adds the term "not" to the positing of mind or body ("mind is not," or "body is not"), at the underlying level of syntax there is only the act of predication itself, the saying of *is* that establishes implicitly the categorical *presence* of what is posited. So the distinction between monism and dualism apparent at the level of content disappears at the subtler level of linguistic form. Here post-Renaissance philosophies are uniformly schismatic.

And *pre*-Renaissance philosophy—what of that? Among the most influential thinkers of the twentieth century is enigmatic philosopher Martin Heidegger. In the introduction to his magnum opus, *Being and Time* (1962), Heidegger demonstrated that modern philosophy essentially *continues* the medieval and ancient traditions. A complete "de-struction of ontology" was called for by Heidegger, ontology being equated with philosophy in general. We are to retrace our steps, work our way back through our twenty-five-hundred-year-old tradition in a manner that would loosen the hold of the structure that has ensnared us. By so returning to the primal origin of philosophy, a fresh perspective would be gained on its fundamental problems, the most crucial of which Heidegger viewed to be the problem of *Being*. According to Heidegger, far from having provided an adequate answer to the question about Being, Western philosophy has never even properly *formulated* it. As a result, this question has been shrouded in obscurity or entirely misunderstood since the time of Plato and Aristotle.

So Heidegger was concerned with the *form* of thinking, over and above its content, and this is closely related to the concern I voiced above in reflecting on the alternatives set forth by Beloff. What limits the form of questioning we have employed from the beginning of the philosophical enterprise? As I proposed at the outset, our manner of questioning presupposes a separation of the questioner from that which is questioned, of the knower from the known, or, we may now say, a splitting off of what is formed (known) from the formative process that gave rise to it. Indeed, this is the essence of *predication*, and it is this that creates dualism in our form of knowing mind and matter, whether or not we make said dualism explicit in the content of our assertions (as Descartes chose to do). By *pre*supposing the division of the known from the knowing process, preconsciously building it into our manner of discourse in such a way that whatever comes to conscious awareness comes only as what is *known* (i.e., as a finished, already processed, form), even when the content of consciousness is itself the "knowing process" or "knower," it will take *form* as merely what is known. Therefore, with "knower" ("subject" or "mind") as well as "known" ("object" or "body") being given in this way, they must be divided from each other, since *anything* known in division from the underlying knowing process (i.e., anything predicated) is itself given as categorically divided.

I venture to suggest that what Heidegger was striving to bring to light in his tentative explorations of Being at its source, is the radical *non*duality of the knower (understood in terms of process, rather than as mere *cogito*) and the known. It is our deeply engrained dualistic mode of operating that downgrades Being, makes it into a mere *being* (a known form), as Heidegger might say. In approaching the question of Being from within the prevailing structure of language and thought, it is compellingly natural for us to predicate Being, to say that "Being is. . ." or that "Being is not. . ." But this predicative manner of addressing Being is just what uproots it, distances it, divests it of its processual dynamism. By predication, Being is circumscribed, reduced to an item at hand, a fixed object of study, a thing to be known that is divorced from the knowing process, a thing thus itself divided from other such things. Of course, Heidegger himself was a product of the philosophical tradition he was reflecting on at this fundamental level, so it was necessary for him to make the attempt to "grope his way out" of said tradition, as translator Joan Stambaugh (1972, p. x) has put it. By the same token, the notorious difficulty in comprehending Heidegger reported by many stems, in part, from the fact that most readers are even more entrenched in the long-held tradition. They struggle to grasp what Heidegger was predicating when, in point of fact, Heidegger—through various unconventional and often poetic styles of expression—was seeking to call to his readers' attention the basic limitation of the predicative mode in confronting the question of Being. In terms of our own immediate concern, the point would seem to be that if we are to surmount the mind-matter split, mind and matter no longer can be predicated as mere modes of being (finished forms), but must be grasped as aspects of Being, i.e., apprehended more primordially, from the prepredicative unity of known and knower (form and formative process).

Nowadays, the professional philosopher typically is a specialist working within a narrowly defined disciplinary area such as symbolic logic or linguistic analysis; as such, his or her concerns are likely to be far removed from those of the student of religion. But when philosophy is practiced in the fashion of a Martin Heidegger, its original scope becomes evident. Issues of ultimate meaning are addressed that are intimately linked to the central issues of religion. The answers that evolved in classical philosophy to these ultimate questions have their counterparts in theology. To demonstrate, I present an adaptation of philosopher Alan Anderson's (1981) pictorial summary of basic theological positions (fig. 11.1).

For the present, let us consider only the first three columns of figure 11.1. Here the space-time reality of differentiated physical process is portrayed by arrows pointing in different directions, while psychospiritual reality appears in unadulterated form as an unfilled, distinctionless circle. It is clear that Anderson's characterization of atheism, traditional theism, and pantheism can be taken as equally representative of materialism, dualism, and idealism, respectively (the justification for treating Cartesian and physicalistic varieties of dualism as equivalent is implicit in what is said of them above). Therefore, Anderson's original diagram has been expanded to a second row incorporating the philosophical correlates.

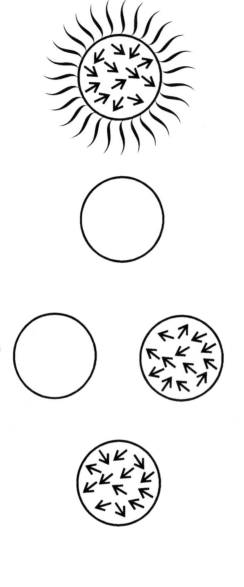

Figure 11.1

Theology	Atheism	Traditional Theism	Pantheism	Panentheism
Philosophy	Materialism	Dualism	Idealism	Non-Dual Duality or Dualistic Monism

Pictorial summary of basic theological positions and their philosophical counterparts
Adapted from Anderson 1981

In the theological counterparts of materialism and idealism, we again can observe the dualism of form that underlies monistic content. In explicitly denying a simple spiritual reality ("spirit is not"), the atheist predicates it implicitly ("spirit *is*"); the pantheist does the same with respect to physical reality. To appreciate fully the pervasiveness of this deeper order of dualism in human culture, let us examine the ancient strain of pantheism that arose in Eastern philosophy/religion.

In the Vedantist tradition of Indian religion, the distinction is made between the finite world of diversity given to the senses and a pristine reality beyond any kind of sensing or knowing, a seamless, utterly undifferentiated totality. We attach ourselves to myriad things, people, and particular ideas and so become enmeshed in *maya*, "illusion." Such fixations on the finite are what keep us in bondage. The challenge to be met is that of *dhyana*: to polish the mirror of consciousness, to gradually purify the field of awareness through unrelenting practice of proper meditation. All things must be surrendered—especially the narrow and distorted view of ourselves; every attachment must be allowed to dissolve until not a single blemish remains to cloud the glass of perception. Then, when this aim has been achieved, we shall see into ourselves with impeccable clarity, see into Self, our true nature, and gain Nirvana.

Evidently, to posit infinite totality in this fashion, to project it as a goal toward which one advances through meditative technique, is to engage in something of a contradiction. For doesn't the positing of a goal and a method for its attainment imply one's current *separation* from that goal? But a state from which we can be separated is one that is bounded and therefore, by nature, *finite*. An authentic infinity would have no boundary whatsoever.

In actuality, *any* predication of the infinite must render it finite, for the act of predicating is an act of circumscribing, fixing a boundary. The problem of symbolizing the infinite lies not in identifying infinity's true characteristics but in the symbolizing process itself; it is a problem of form rather than content. As long as one adopts the subject/predicate form of expression, the deeper dualism will be working implicitly to undermine the monism asserted at the level of content.

What we are seeing is that predicating the infinite brings us true monism no more effectively than predicating the "infinitesimal"—that is, attempting to *reduce* one member of the spirit-matter (or mind-matter) pair to nothing or zero, to negate it by asserting that it "is not" (as discussed above). In Eastern pantheism, both forms of predication are in evidence. Spiritual unity is affirmed as ultimate while material diversity is dismissed as *maya*, as mere "illusion." However, though spirit may be elevated to the infinite and matter reduced to zero in *what* is said about them, in the *form* of speaking (thinking and acting), each is implicitly made *finite*, made into categorically circumscribed spheres.

PARADOX AND THE PRINCIPLE OF NONDUAL DUALITY

Now let us turn to the fourth column of figure 11.1 and the approach known as *panentheism*, a relatively recent development in Western theological thought (see Woods 1981). While the pantheist would hold that God (or spirit or Self) simply is all, the panentheist's conviction is that God is *in* all. Anderson depicted this view by showing psychospiritual emanations radiating from the sphere of material process. This representation may be interpreted as portraying a synthesis or fusion of spirit and matter. We may picture the unfilled circle and the one filled with arrows being superimposed on each other in such a way that a thorough-going unity is conferred (in contrast to dualism); yet this is not a unity in which one realm merely is absorbed by the other (as is the case in the traditional forms of monism). Nor could it be said that in merging, separate spheres lose their distinguishing characteristics by dissolving in some state of simple neutrality. Indeed, according to the theologian Richard Woods (1981, p. 195), unlike conventional theological positions, panentheism is "neither elegant nor simple. It is paradoxical." The fundamental formula for paradox, $X = $ not-X, applies to panentheism. While being entirely distinct, matter and spirit nevertheless interpenetrate completely to share the same identity. In my own previous writing, such a seemingly incongruous relation is termed "nondual duality" *[see chapters 10 and 14]*.

I also have referred to this philosophy of paradox as "dualistic monism" (Rosen 1983), following the lead of cosmologist Nahum Stiskin (1972). Using the Shinto religion of Japan as his point of departure, Stiskin explored an aspect of Eastern philosophy suggestive of a subtlety far greater than one-valued pantheism.

His most pregnant metaphor is that of the "Divine Sword," which is wielded by the August Master of the Center. The August Master dwells

> at the point of the splitting of the unitary energy of life, [and] can be said to wield a divine sword which slices that energy into its two manifestations [i.e., yin and yang] and thereby creates polarity. This deity, however, is the consummate swordsman who, although cutting into two, does so with such speed and precision that the fluid of life continues to flow between the resulting halves. They therefore remain continuous and inter-twined. The two energies of the world . . .are always one in their dialectical interaction. Yet they are distinctly *two*. Or we can say that they are *two-in-One*. (Stiskin 1972, p. 90)

Of course, if this "paradoxical logic of the universe" (Stiskin 1972, p. 21) were seen to derive from a *single* principle of unity independently embodied in the "August Master," then the Shinto philosophy would reduce to pantheism. The question is whether one allows the implication of a boundary dividing the domain of dualistic monism from a realm of absolute totality to which it would be subordinated. Since Stiskin implied a close relationship between Shinto and Zen (the ancient Chinese law of the Tao being central to both), it should be useful to consider the commentary of D. T. Suzuki (1969), who provided a valuable insight into the distinction between two Zen approaches.

Suzuki identified the Northern School of Chinese Zen with the meditational technique of *dhyana*, the essentially pantheistic procedure of gradual purification described above: Sitting in a posture of cross-legged devotion, one aims to polish the mirror of consciousness until every vestige of worldly illusion is erased and one ascends to a plane of unadulterated, totalistic bliss. Recognizing the dualism inherent to this orientation, Suzuki was more favorably disposed toward the Southern School. Here the "technique" was to *renounce* preoccupation with technique, repudiate the idea of a circumscribed goal from which one is separated and must methodically strive to attain. But rather than rejecting *dhyana* outright, the proponents of the Southern School contended that for the meditative experience to be spontaneous and authentic, it must be practiced in a form known as *prajna*— the sudden realization that infinite totality and finite diversity, while being as different as they can be, nevertheless are *one and the same*:

So long as the seeing [into infinite totality] is something to see, it is not the real one; only when the seeing is no-seeing—that is, when the seeing is not a specific act of seeing into a definitely circumscribed state of consciousness—is it the "seeing into one's self-nature." Paradoxically stated, when seeing is no-seeing there is real seeing. . .This is the intuition of the Prajnaparamita. . . It is Prajna which lays its hands on Emptiness, or Suchness, or self-nature. And this laying-hands-on is not what it seems. . . Inasmuch as self-nature is beyond the realm of relativity, its being grasped by Prajna cannot mean a grasping in the ordinary sense. The grasping must be no-grasping, a paradoxical statement which is inevitable. To use Buddhist terminology, this grasping is accomplished by non-discrimination; that is, by *non-discriminating discrimination*. (Suzuki 1969, pp. 28-29 and p. 60; emphasis added)

The last phrase of the quotation from Suzuki is emphasized to call attention to the expression of nondual duality or dualistic monism. In the proper practice of Zen, "all the logical and psychological pedestals which have been given to one are now swept from underneath one's feet and one has nowhere to stand" (Suzuki 1969, p. 26). This means that at every turn one must resist the temptation to fall back on the conventional, dualistic mode of operating. Paradox must be allowed to pervade. To summarize what I am proposing, panentheism in the West and Zen in the East each offers a nondually dual alternative to the established, dualistic order of religion and philosophy in its respective culture—the alternative of paradox.

But is this *all* that can be said? The dictionary defines the paradoxical as that which is absurd, enigmatic, or contradictory. In asserting that "*X* is not-*X*," the conventional, subject/predicate format *is* being used, but in a manner in which it denies itself, for the content it expresses calls this form into question. The paradoxical statement amounts to a declaration that the syntactical boundary condition that would delimit *X* cannot effectively do so; predicative boundary assignment is confounded so that even though *X* implicitly is being posited as distinct from that which is external to it, at the same time, it is inseparable. To be sure, this does give voice to the principle of nondual duality. Yet because the traditionally *dualistic* form of expression is being employed, the principle is being stated essentially in a negative fashion. To assert that "*X* is not-*X*" is to imply that the format one is using cannot fully convey the meaning intended. A sense of tension is created between the predicative format and a content that flies in the face of simple predication, suggesting its inadequacy. Then must we rest content with the acknowledgment that nondual duality is "merely" paradoxical, or can we find a way to give positive voice to it?

EMBODYING NONDUAL DUALITY

The question at hand is how form of expression relates to content expressed. This is an issue of central importance in the field of *artistic* expression, one that is crucial enough in the present context to warrant a detailed illustration (see Rosen 1981). A graphic example is found in the art of M. C. Escher. *Tower of Babel* (fig. 11.2) is one of Escher's earliest works. He explained it as follows: "On the assumption that the period of language confusion coincided with the emergence of different races, some of the building workers are white and others black. The work is at a standstill because they are no longer able to understand each other" (Escher 1971, p. 9).

Commenting generally on his own efforts in this early period, Escher said: "[The prints] display no unity as far as their subject matter is concerned. They are all representations of *observed* [i.e., external] reality" (Escher 1971, p. 9; emphasis added). And with specific regard to *Tower of Babel*, Escher concluded that "it was not until twenty years later that this problem was thoroughly thought out (see. . .'Other World'. . .)" (Escher 1971, p. 9). Upon inspecting *Other World* (fig. 11.3), we see the content that Escher actually had been trying to express: not racial or linguistic separation but separation in and of itself—radical disjunction. In this later print both vertical and horizontal perspective are used to create three separate planes of existence. With the theme of disjunction thus *embodied in the very geometry of the work*, it is no longer necessary to symbolize it in a merely external fashion (i.e., to indicate the theme by making reference to external reality). And now the observer of the work need not *infer* the subject matter from outside knowledge s/he may have gained about the referents of the symbol (i.e., that the color disparity of the construction workers signifies racial divergence presumed to be correlated with a linguistic divergence that has made communication impossible, thereby creating the effect of a profound separation). In *Other World*, the content that had been latent

Figure 11.2

Tower of Babel

Figure 11.3

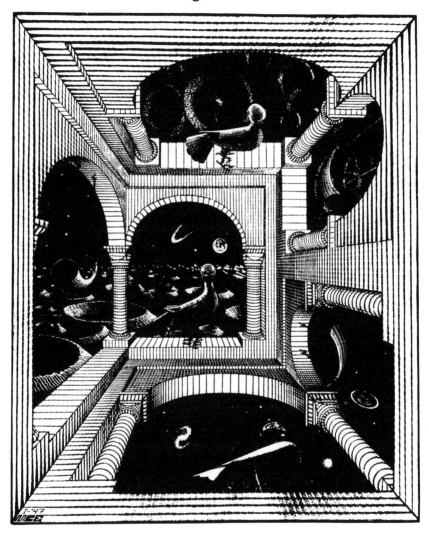

Other World
© 1947 M. C. Escher / Cordon Art - Baarn - Holland. Collection Vorpal
Gallery New York (Soho) / San Francisco.

is made manifest in the form. The print therefore possesses
greater unity of expression than its predecessor. The gap between
the intended and the given has been closed in a natural manner,

from within, thus obviating the necessity of constructing artificial bridges.

The subject matter presently requiring expression is that of *nondual duality*. In the previous section we saw how conventional linguistic representation can voice this theme in but a limited, negative way, as the mere *predication* of paradox (the assertion that "*X* is not-*X*"). What of the *pictorial* representation shown in figure 11.1, column 4? I submit that Anderson's diagram has the same status as Escher's *Tower of Babel*: It symbolizes its intended subject matter without embodying it. One must *infer* the idea of 'nondual duality' from externally gained information about the meaning of the emanations issuing from the arrow-filled circle. The principle is not inherent to the structure of the depiction . . .

[At this juncture, the Moebius principle is systematically brought into play via our three "bodies of paradox"—the Necker cube, the Moebius strip and the Klein bottle—so as to give positive expression to nondual duality. Our concrete embodiments are seen to serve the purpose of fleshing out the alternative to the philosophico-religious tradition of dualism and, at the same time, of accommodating the phenomenon of psi, since "hybrid" psi reality is construed as entailing the paradoxical interpenetration of physical and nonphysical realities.

However, it is acknowledged that the three "Moebial structures" have their own limitations as embodiments of paradox. This is brought out in presenting the Necker cube, Moebius strip and Klein bottle in the form of a dimensional progression, with each structure approximating the higher-dimensional nondual duality more closely than its antecedent, but none embodying it completely. Thus, while the two-dimensional integration of Necker-cube perspectives certainly captures nondual duality in a more concrete, intrinsically coherent way than the Anderson diagram or the mere predication of paradox, it is recognized that the experience with the cube is merely perceptual; as such, it is limited to the province of physical sensation. Necker cube phenomenology cannot directly engage the added dimension of the psyche, cannot fully embody the nondual duality of physical and nonphysical realms that psi would entail. Therefore, with respect to psi, the cube remains merely symbolic. A deeper embodiment evidently is necessary.

Next, to pave the way for this, the nondually dual structure is tangibly "materialized" via the Moebius strip.]

In the surface of Moebius, nondual duality is embodied more concretely than in the experience of perspectival integration performed with the Necker cube. Whereas the cube is but a two-dimensional visual representation of higher-dimensional reality, the Moebius strip engages the third dimension. So the one-sidedness suggested by viewing the cube in the special way is more tangibly delivered in the Moebius. And yet the band of Moebius obviously does not go beyond three dimensions. Consequently, it is wholly contained in the realm of physical sensation, constrained by *physis* no less than the Necker cube. Since the duality bridged by psi would include the *para*physical, the Moebius cannot embody it; like the cube, it can only *symbolize* the interpenetration of *physis* and *psyche*. Still another step must be taken.

An interesting feature of the Moebius band is its asymmetry. Unlike an ordinary cylindrical ring *[fig. 1.2a]*, a Moebius surface has a definite orientation in space; it will be produced in either a left- or right-handed form. If both a left- and right-facing Moebius were constructed and then "glued together," superimposed on one another point for point, the topological structure known as the "Klein bottle" would result.

The Klein bottle is the higher-order counterpart of the Moebius strip. It has the same property of asymmetric one-sidedness as the Moebius, but while the Moebius twist entails a projection into a third dimension, the Klein bottle would project into a *fourth*. For this reason, the production of a proper physical model of the bottle cannot be completed. Left- and right-facing Moebius bands cannot be superimposed on each other in three-dimensional space without tearing the surfaces.

There is a different but topologically equivalent way to describe the making of a Klein bottle that should be quite revealing for our purposes. Let us approach this by way of a comparison. Both rows of figure 11.4 depict the progressive closing of a tubular surface that initially is open. In row one, the end circles of the tube are joined in the conventional way, brought together through the three-dimensional space outside the body of the tube to produce the doughnut-shaped form technically known as the "torus" (the higher-order analogue of the cylindrical ring). By contrast, the end circles of row two are superimposed

Figure 11.4

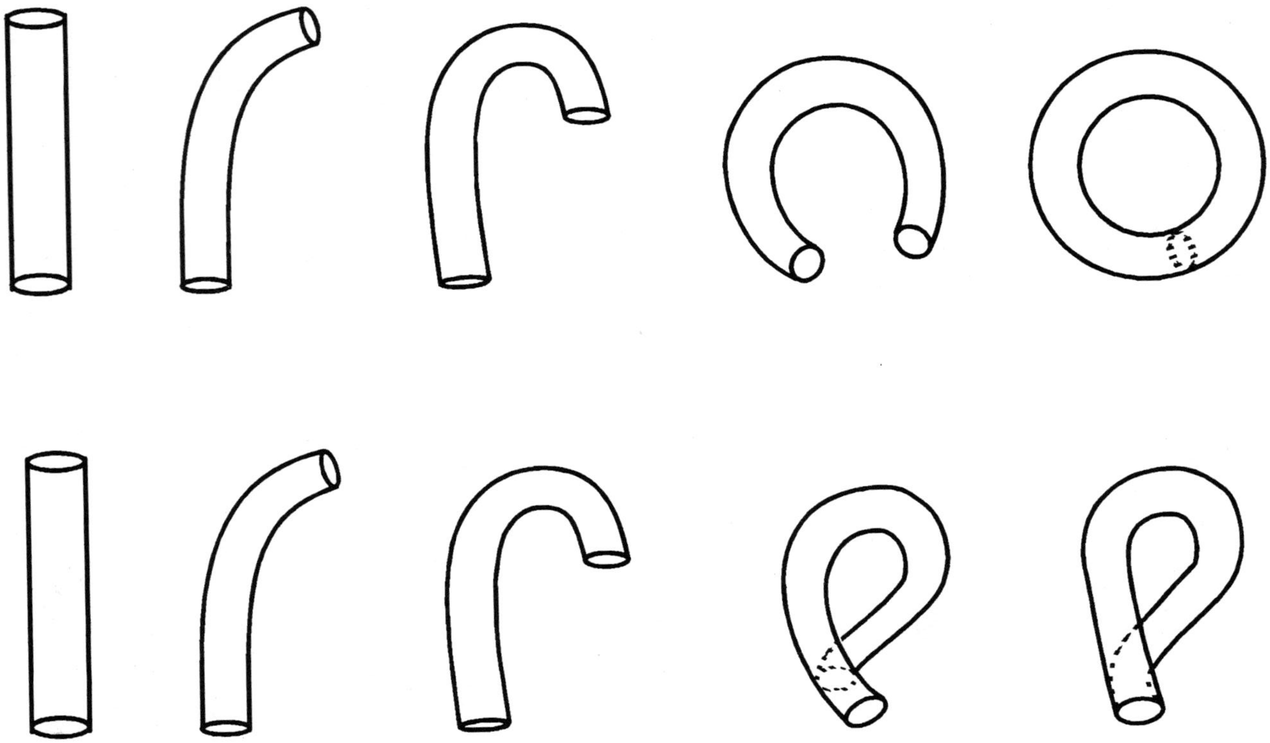

Construction of torus (upper row) and Klein bottle (lower row)

from *inside* the body of the tube, an operation requiring the tube to pass *through* itself. This results in the formation of the inside-out Klein bottle. Indeed, if the structure so produced were bisected, right- and left-oriented Moebius bands would be yielded. But in three-dimensional space, no structure can penetrate itself without cutting a hole in its surface, and topologically, this is impermissible. So from a second perspective we see that the construction of a Klein bottle cannot effectively be executed when one is limited to physical dimensionality. Mathematicians are aware that a form that penetrates itself in a given number of dimensions can be produced without the prohibited cutting if an *added* dimension is available (for a demonstration of this, see Rucker 1977). In the case of the Klein bottle, the additional dimension needed for full and proper expression is the paraphysical one.[1]

The fact that the Klein bottle implicates a fourth dimension clearly does not require us to regard it as *wholly* nonphysical, as purely a product of *psyche*. Note the wording I employed above. I did not make the claim that the Klein bottle *is* four dimensional but only that it *projects* into a fourth dimension. This distinction is critically important and may be appreciated by considering the *tesseract*, a mathematical entity frequently used as an example of a four-dimensional structure. The tesseract is taken as an "imaginary" extrapolation of a three-dimensional cube (not to be confused with the Necker cube) to four spatial coordinates. Each of the "faces" of this hyper-cube is itself a three-dimensional cube defining the lower limit or boundary condition of the higher-dimensional structure. Were a "face" to be viewed from the three-dimensional vantage point, *only* a cube would be experienced, since a tesseract "face" is a closed symmetric form, complete in itself and totally indistinguishable from any ordinary physical cube. Not the slightest hint would be given of paraphysical extension. Succinctly put, the tesseract is a simply "nonphysical" ("imaginary") entity with a simply physical boundary condition.

I suggest that the Klein bottle, in marked contrast, embodies the nondual duality of *psyche* and *physis* we have been seeking to articulate. Here the working of the paraphysical is encountered in the *midst* of the physical, in its very "incompleteness." The objectionable hole that would be necessary to finish construction of the Klein bottle in three dimensions can be said to result from attempting artificially to contain a form that flows unbrokenly into the fourth, inner dimension. In flowing through itself, the inside-out Klein bottle flows *between* dimensions, fluidly bridges the gap between *physis* and *psyche*. While the tesseract strictly upholds the categorical separation of the physical (observable) and paraphysical ("imaginary"), the Klein bottle transcends this dualistic state of boundedness. Its self-penetration reflects the hybrid quality construed as fundamental to the nature of psi phenomena, the paradoxical relation whose positive expression has been the central aim of this essay.

Naturally, to comprehend the philosophical perspective I have offered is not to experience psi connectedness in a literal manner. Though the Klein bottle representation of psi may be intuitively quite compelling, it does not add up to psi itself. For that, it seems we would need to go beyond thought or intuition alone and include a concrete *feeling*, a palpable *sensing* of nondual duality. Just as the Necker cube permits us to experience nondual duality at a perceptual level when we *enter into* the cube in the appropriate way, we would have to enter into the Klein bottle at *its* level, embody this embodiment with our very own bodies. I propose that only through such an act of embodying would psi itself be lived. Nevertheless, the intuitive grasping of nondual duality should be a significant step. As I have shown in this exposition, the mere idea of psi has been rejected for centuries because of its "counterintuitive" quality—that is, its underlying incompatibility with the prevailing philosophico-religious framework of dualism that has exerted such a powerful influence on us all, parapsychologists not exempted.

I end my presentation with a quote from Zen philosopher D. T. Suzuki:

> A monk asked Li-Shan: "All things return to Emptiness, but where does Emptiness return?"
> Li-Shan: "The mouth is unable to locate it."
> Monk: "Why not?"
> Li-Shan: "Because of the oneness of inside and outside." (1969, p. 93)

PART III.

Dialogues with David Bohm

INTRODUCTION

The material in this section covers the period of my writing from 1982 to 1984 and is devoted to my essays on and correspondence with the physicist and philosopher I have cited a number of times throughout this book: David Bohm. Until his recent death, Bohm was the most prominent member of a small group of scientists dedicated to addressing the question of wholeness (healing, transpersonal realization) in a serious way. He attempted to go beyond general programmatic statements about the relationship between wholeness and modern physics to a detailed, systematic exploration. I know of no individual whose influence has been greater in fields that are of central concern to this volume.

The first selection (chapter 12) is an essay offering an interpretation of Bohm's seminal work, *Wholeness and the Implicate Order* (1980). My paper was written both to celebrate Bohm's important contribution and to enter into dialogue with it. In the summer of 1982, I sent a preprint of the piece to Bohm, and this initiated a correspondence between us that lasted around a year and a half. Our interchange reached its peak over the six-month span from June to November 1983. Chapter 13 contains verbatim transcripts of these discussions, published with Bohm's permission (while the substance of our dialogue appears in its original form, I have deleted some peripheral allusions and have added some references). Note that for Bohm's first letter to me of this period (written in June 1983) and for his subsequent two-part letter (of August 1983), I followed the procedure of transcribing his handwritten prose into type and inserting my own remarks in double brackets as I went along. The results were then sent back to him for his own response and further comment.

Following upon our exchange of letters, in 1984 I was invited to participate in "Physics and the Ultimate Significance of Time," a conference held under the auspices of the Center for Process Studies, a group devoted to studying the philosophy of A. N.

Whitehead. My role was to offer a response to Bohm's conference paper. The essay that was produced (chapter 14) focuses upon the paradox of time and, again, is both an appreciation of Bohm's work and a dialogue with it.

12

David Bohm's *Wholeness and the Implicate Order:* An Interpretive Essay (1982)

We live in an era of uncertainty and alienation, a time when all of our standards, our most hallowed institutions and cherished beliefs, have been opened to question. As we enter the final years of the millenium, the disintegrative process appears to be accelerating, building to a crescendo: technology run amok, bureaucratic breakdown, ecological crisis, international banditry and terrorism, suicidal militarism, nuclear insanity—a new disaster seems to threaten us at every turn. This devastating trend toward fragmentation has led many to be despondent, to surrender to a sense of futility. Yet there are those who are able to perceive our current dilemma as a challenge to be met.

Among the serious thinkers and visionaries who have sought a positive course of action, the physicist David Bohm is a leading figure. Bohm has been striving to bring into harmony the contemporary knowledge of our post-Einsteinian world and the perennial wisdom of the ages. His hope is that such a reconciliation might serve to catalyze the constructive transformation of our beleaguered species. In the past few years, the efforts of Professor Bohm have attracted wide attention and been acknowledged (e.g., *Brain/ Mind Bulletin* 1977; *Re-Vision* 1978; Weber 1981; Pearce 1981).

Recently, Bohm himself has attempted to integrate decades of reflection into a comprehensive statement. From this labor comes *Wholeness and the Implicate Order* (1980), the book upon which the present essay is a meditation. Whether or not Professor Bohm can accept all my interpretations of, and elaboration upon, his concepts and perspectives, I am deeply indebted to him for providing the impetus and means for my exploration.

MODERN PHYSICS AS AN INDICATION OF WHOLENESS

Naturally, as a theoretical physicist, Bohm is greatly concerned with the current status of his field. Yet he is first and foremost a *human being*. Therefore, his *greatest* concern is not with physics as such but with our problem of fragmentation and search for wholeness.

201

Chapter 1 begins with the observation that "fragmentation is now very widespread, not only throughout society, but also in each individual" (Bohm 1980, p. 1). The fragmentation in our lives is seen to stem from a fragmentary way of thinking that controls our perceptions and governs our actions. Early in human history, divisions were created chiefly to serve practical ends (e.g., land to be farmed was divided into separate fields where different crops could be grown). But the notion of dividedness subsequently became a philosophical absolute, a belief system telling us that, at bottom, our world *really is* composed of segments in a state of disunion. It is the persistence of this underlying vision that

> has led to the growing series of extremely urgent crises that is confronting us today. . . pollution, destruction of the balance of nature, over-population, world-wide economic and political disorder, and the creation of an overall environment that is neither physically nor mentally healthy for most of the people who have to live in it. Individually there has developed a widespread feeling of helplessness and despair, in the face of what seems to be an overwhelming mass of disparate social forces, going beyond the control and even the comprehension of the human beings who are caught up in it. (p. 2)

By so identifying the dilemma presently facing humankind and demonstrating its basis in a fragmentizing worldview, Bohm establishes the primary motive for his efforts: to help promote the rediscovery of *wholeness.* "To be healthy is to be whole," notes Bohm, pointing out that "the word 'health' in English is based on the Anglo-Saxon word 'hale' meaning 'whole'" (p. 3). It is Bohm's conviction that his own field of study, modern physics, can play a key role in the quest for wholeness. By properly developing the implications of this fundamental and still emerging discipline, a therapeutic transformation of thinking, perceiving, and acting can result.

Indeed, a central role has been played by *classical* physics— not in facilitating wholeness but in fostering the opposite. While post-Renaissance figures like Galileo, Descartes, and Newton were giants in their time, the effect of their work was to powerfully reinforce the outlook of fragmentation. In contributing mightily to our ability to analyze, thus control, our material universe, they gave impetus to a mechanistic way of ordering reality that continues to exert a pervasive influence upon us. Bohm summarizes the mechanistic order in classical physics by describing its principal feature:

> The world is regarded as constituted of entities which are *outside of each other*, in the sense that they exist independently in different regions of space (and time) and interact through forces that do not bring about any changes in their essential natures. The machine gives a typical illustration of such a system of order. Each part is formed (e.g., by stamping or casting) independently of the others, and interacts with the other parts only through some kind of external contact. (p. 173)

The contrasting, *non*mechanistic character of *modern* physics is typified for Bohm in the quality of a living organism: "Each part grows in the context of the whole, so that it does not exist independently, nor can it be said that it merely 'interacts' with others, without itself being essentially affected in this relationship" (p. 173).

According to Bohm, "the theory of relativity was the first significant indication in physics of the need to question the mechanistic order" (p. 173). In Einstein's conception of the universe, we can no longer speak properly of rigid bodies, entities with absolutely definable boundaries; instead we must regard these as locally intense concentrations of an overall field, singularities that ultimately merge into the totality without ever actually establishing themselves as completely autonomous units. So, with Einstein, "we come to an order that is radically different from that of Galileo and Newton—the order of *undivided wholeness*" (p. 125).

The challenge to the mechanistic order is even more evident in *quantum physics*, the study of atomic and subatomic processes. Time and again Bohm calls attention to its central findings, which are entirely unexpected from the familiar perspective of macroscopic observation and are basically suggestive of wholeness. Wholeness is implied in the *indivisibility of motion* at the quantum level (thus an electron can go from one state to another without passing through any states in between). It is indicated by the *nonlocal connectedness* of quantum elements (e.g., a pair of electrons that are initially associated, then separated by a vast space-time interval, may nevertheless behave as if still linked in an immediate and intimate way). Most profoundly perhaps, quantum level wholeness expresses itself through the "problem of measurement" in which events observed cannot rightly be regarded as occurring independently of the observation process. We may take this to mean that the observer/scientist is no mere passive witness to what is "out there" but is, in a sense, an active participant in the *creation* of that "external" reality. Hence, not only is wholeness an inherent property of the phenomena recorded but the recorder him- or herself is included as an aspect of this wholeness.

From the "quantum leap" (indivisibility of motion) and the "quantum connectedness" (nonlocalness) the fundamental lesson to be derived is that microprocesses transpire in blithe indifference to the proscriptions of space. Space, as traditionally construed, is the premier principle behind fragmentation in classical physics.[1] To conceive of entities as situated in space is to conceive of them as *necessarily* segregated, for space is taken as an extended continuum of infinite density, that is, as composed of an infinite number of tightly packed but absolutely distinct point locations. Here, localness is mandated by the presupposition that every point is completely external to every other point. This being the case, interaction can occur *only* by continuous, *non*indivisible passage from one separate point to its abutting neighbor. Indeed, Western science's entire program of classical analysis (the basic tool of which is the calculus invented by Leibniz and Newton) depends on the assumption that space is infinitely *divisible*. So it is our idea of space that undergirds our mechanistic way of ordering reality, as Bohm implies. Whatever the particular content of a dynamic occurrence, be it the orbital action of a planet or the act of making love, its mechanical character is guaranteed when it is described in terms of our notion of simply continuous space. By calling into question the differentiative, fragmentizing manner of viewing space, modern physics contributes to indicating wholeness.

But it is clear that *fragmentation* is still the order of the day. Then what must we do to further develop what is now little more than a promising portent? What shall our program be? This is the paramount issue of Bohm's exposition. Granting that the mechanistic order of classical physics bears substantial responsibility for our extreme fragmentation, and that the phenomena of modern physics signify wholeness, surely we need to articulate a new, *non*mechanistic approach that does full justice to these phenomena. Bohm's sincere belief is that the proper formulation of physics would be a great achievement not merely for the community of theoretical physicists but for the human community at large. But how specifically can we move toward this goal?

THE PERSISTENCE OF MECHANISTIC THINKING
IN CONTEMPORARY PHYSICS

Bohm notes early on that the prevailing attitude among contemporary physicists is "much against any sort of view giving primacy to...undivided wholeness" (p. 14). The essentially mechanistic reaction to nonmechanistic indications may be said to take two main forms: *implicit* and *explicit*. The former predominates and is associated with the Copenhagen interpretation of quantum physics, or, what Stapp calls the "pragmatic approach":

> According to the pragmatic approach quantum theory should be viewed as merely a set of rules for calculating correlations among observations... One should refrain from making ontological assumptions about the nature of the world that "lies behind" the observations. The observations themselves, together with their connections and correlations, are what is real for us. The construction of ontologies (theories of what exists) lies outside the scope of science. (Stapp 1979, p. 9)

Thus advocates of the "pragmatic" approach would create the impression that the calculational system developed to deal with quantum processes is a pure abstraction devoid of interpretive content. Yet this appears contrary to fact. Far from being interpretively neutral, the mathematical structure of extant quantum theory rests upon a definite assumptive base and the ground selected is the *classical* one. To follow the standard analytic operating procedure, it is tacitly presupposed that the quantum domain possesses infinite divisibility, that it is a domain of extended, simply continuous spaces. Hence, the mechanistic ontology operates after all, though on an implicit level. (Indeed, the philosopher well might ask the quantum mechanical "pragmatist" whether it is ever possible to separate completely our method of knowing the world from our assumptions about the world that is known.)

The *explicitly* mechanistic response to quantum wholeness is distinguished only by the fact that no attempt is made to deny the ontological consequences of the formalism. Instead, they are accepted and advanced. Stapp's generic term for this frankly ontological approach is the "absolute psi interpretation." "Psi" refers to the basic mathematical wave function of quantum physics, described as a vector in Hilbert space. In the "absolute psi interpretation" (which includes the view that the psi wave collapses upon observation and the alternative, "many worlds" view that it does not), the mathematical wave is seen as a veridical representation of the underlying quantum reality. For us the relevant point is that whether a "pragmatic" or "ontological" stance is adopted, the formalism remains invariant. It is this formalism, which results from the attempt to extrapolate the simple continuity assumption to the quantum level, that will now be examined more closely.

Bohm opens his chapter on "quantum theory as an indication of a new order in physics" (chapter 5) by describing the officially sanctioned response given when the order of *ancient Greece* became suspect. That view of the world was predicated on the idea of the circle, cosmologically expressed in the conviction that celestial bodies trace perfectly circular orbits:

> To be sure, when more detailed observations were made on the planets, it was found that their orbits are not actually perfect circles, but this fact was accommodated within the prevailing notions of order by considering the orbits of planets as a superposition of *epicycles*, i.e., circles within circles. Thus, one sees an example of the remarkable capacity for *adaptation* within a given notion of order, adaptation that enables one to go on perceiving and talking in terms of essentially fixed notions of this kind in spite of factual evidence that might at first sight seem to necessitate a thorough-going change in such notions. (p. 112)

I might add that, in the notion of "epicycles," there was a kind of *mockery* of the image of the circle, an extravagantly complex, wholly gratuitous replication of this image, one that did not *truly* achieve what it sought, since the circle actually had lost its original meaning and effectuality. Yet those bent on preserving circularity blinded themselves to this.

Of course, Bohm's chief concern is not with the ancient order of Greece but with our current perseveration upon the mechanistic order of post-Renaissance classical physics, which I have suggested is epitomized in the concept of simple continuity. While microphysics does signify *dis*continuity, the *formalism* that has been applied to it is contrived to preserve continuity. The rather arbitrary and unparsimonious character of this accommodative manipulation may best be seen by reconsidering the quantum nonlocalness from a somewhat broader perspective. Nonlocalness is implied not only in the direct linkage of spatially remote particles but in the fundamental inability to determinately fix the position of a given particle. Since the position of a system in a simply extended, infinitely differentiable continuum is always uniquely determinable, the basic quantum *in*determinacy strongly suggests that *microworld* systems *do not occupy* distinct positions, that is, that microspace is *not* simply continuous.[2] But discontinuity is precisely the quality that is anathema to mechanistic thinking. Therefore, in order to bar the appearance of the radically nonclassical discontinuity, a *multiplicity* of simply continuous "spaces" is axiomatically invoked to account for the "probable" positions of the particle—"it" is locally "here" with a certain probability or "there" with another. This collection of subspaces constitutes the Hilbert space, the accepted formalism of quantum mechanics. Yet we may ask whether continuity is *truly* preserved by the Hilbert space device.

Each subspace of the multispace expression is made simply continuous within itself to uphold the mutual exclusiveness of the alternative positions of the particle. Such subspaces must be *disjoint* with respect to *each other*, their unity being imposed externally, by fiat, rather than being of an internal, intuitively compelling order. Thus, in the name of maintaining mathematical continuity, a rather extravagantly *dis*continuous state of affairs is actually permitted in the standard formalism for quantum physics, an indefinitely large aggregate of essentially discrete, disunited spaces. Do not the self-continuous subspaces of the Hilbert formalism mock simple continuity in the very same way the epicycles of antiquity mocked circularity? How explicit must the analogy be made before it is recognized that the ancient folly is being played out again in our modern context?

Bohm's own treatment of the continuity problem begins from the standpoint of physics' long-held aspiration to incorporate the quantum and relativity theories in a single account. Coherent unification is blocked by the inability to absorb the intrinsic quantum discontinuity into relativity's continuous field expressions.

When the attempt to do so is made, unmanageable divergences (infinities) arise in the calculations. Bohm goes on to identify some of the "corrective measures" that are taken to remove the divergences (e.g., renormalization, S-matrices, etc.), methods that suffer from a self-serving, arbitrary character "reminiscent of the way in which the Ptolemaic epicycles could be made to accommodate almost any observational data" (p. 134). The author concludes that because of this arbitrariness and "what at least appear to be some serious contradictions" in the theory, the relativistic quantum field theory "at bottom . . . remains generally unsatisfactory" (p. 134).

Next, Bohm observes that even when we do not try to relativize it, the quantum theory "contains an implicit attachment to a certain very abstract kind of analysis, which does not harmonize with . . . indivisible wholeness" (p. 137). The difficulty is attributed to the essential *linearity* of the Hilbert space expression for quantum physics, and this accords with the way of viewing the problem I suggested above, as we may see from the simplest definition of linearity: the quality of being "*extended like a line*" (Mandel 1969, p. 193; emphasis added). An extended space, whether geometric or algebraic, is one whose elements do not overlap: they are distinct, mutually external, though they may be juxtaposed. In other words, an extended or linear space has the basic character of *simple continuity*. (In the case of linear or continuous algebraic "spaces," distinct "points" are algebraic terms or transformations that interact according to given rules of operation but basically remain unaltered.) In summarizing the linear strategy, Bohm asserts that "this whole line of approach reestablishes at the abstract level of statistical potentialities the same kind of analysis into separate and autonomous components in interaction that is denied at the more concrete level of individual objects" (p. 138). The point, of course, is that quantum phenomena, indicative as they are of indivisible wholeness, *cannot* be convincingly represented in such fragmentizing terms. It is by making the effort to do this that we are led to questionable accommodations of the Ptolemaic sort.

We have seen that, just as the epicycle stratagem fails to uphold authentically the idea of the circle, the "epi-continuous" Hilbert space fails with continuity. The quantum discontinuity, being essentially inescapable, will make its presence felt in one way or another. If denied positive expression (as it is in the Hilbert-space approach), it will be manifested *negatively*, in the very arbitrariness and extravagance of the brute force denial procedure. Then it appears we have no choice but to *accept* the discontinuity, however difficult that may be. Indivisible wholeness no longer can be ignored, nor can we merely pay lip service to it by introducing it at one level only to deny it at a level more basic. In recognition of this, Bohm calls for a radical reformulation of physics: "one may consider the notion that through a thorough-going non-linearity, or in some other way, quantum theory may be allowed to change, so that the resulting new theory will. . .imply undivided wholeness" (p. 138). The uniqueness of physicist David Bohm, the key quality that distinguishes him from the vast majority of his colleagues (whatever their specific programs), is his understanding that, somehow, we must take wholeness as *primary*, as the "place" from which to *begin*.

WHOLENESS AND THE REPRESENTATION OF IMPLICATE ORDER

The Analogy of Enfoldedness

For Bohm, making a new beginning means finding a fundamentally new way to order our thinking. To this end, the notion of order itself is examined, the author demonstrating that it is not limited to causal/deterministic (i.e., mechanistic) relations but includes a more general form of reasoning, that of analogy: " 'As things are related in a certain idea or concept, so they are related in fact' " (p. 114).

It seems there is a simple analogy behind the whole Western mechanistic enterprise. All of classical physics and mathematics—including Euclid's geometry, the Cartesian system of coordinates, Leibniz's and Newton's analytic calculus—is based on the idea of *explicateness* or *unfoldedness*: "As things are related in an explicate or unfolded spatial arrangement, so they are related in fact." For Bohm, the mechanistic order is an *explicate* or *unfolded* order. In the notion of unfoldedness, we readily recognize the concept of extension or linearity, of mathematical continuity. We can grasp this relationship instantly and in the simplest terms by using a blank sheet of paper as a model of space and marking upon this sheet, say, two separate points at opposite ends. It is obvious that the points will remain separate and distinct from one another as long as the paper is not folded (along the axis perpendicular to the axis on which the points are aligned). The unfolded paper therefore constitutes an explicate ordering of the points.

Now Bohm recommends that we may begin to approach the undivided wholeness signified by modern physics if we change the underlying analogy that governs our thinking, switch from an explicate to an *implicate* or *enfolded* form of order. To gain an "immediate perceptual insight into what can be meant by undivided wholeness" (p. 145), he offers the example of the hologram, a type of photograph "each part [of which] contains information about the *whole object* . . . [so that] the form and structure of the entire object may be said to be *enfolded* within each region of the photographic record" (p. 177). A second analogue of enfoldedness that Bohm develops at length concerns the stirring of initially discrete droplets of ink into a viscous fluid, an operation causing them to "intermingle and interpenetrate" each other throughout the glass apparatus in which they are contained. As an alternative, the notion of enfoldedness may be delivered by the simple act of folding the above-mentioned sheet of paper so that the spatially segregated points overlap. And if this paper is folded upon itself in a more complete fashion, compressed into a small ball, for instance, the idea of the interpenetration of *all* explicately distinct spatial elements may be conveyed.

Can the Form of Representation Be Improved?

Undeniably, the paper-folding model and Bohm's effective holographic and ink-in-fluid analogies call attention to wholeness. But, as Bohm implies when he characterizes the latter example as "*only* an analogy" (p. 179), this form of representation is limited. Analogical order entails both similarity and *difference*. In the models we have discussed, certain features do indicate wholeness, while others do *not*. In fact, the *process* by which each of the models expresses wholeness is basically *mechanistic*, because each involves a continuous transformation *in* space carried out by an external agent (folding the paper, stirring the fluid, forming the hologram); in contrast, *quantum*-level processes constitute a *self*-transformation *of* space, one in which spatial continuity is *lost*, spatial separation nullified. Thus, when I fold a sheet of paper in our familiar, simply continuous macroworld space, the idea of intimate point connectedness surely is *suggested*, yet in actuality these points are merely being juxtaposed—like the parts of a machine, they remain essentially extrinsic to each other. Their *literal* interpenetration would require the deeper order of connectedness found in the microworld. The same is true of the act of stirring the viscous fluid. The droplets of ink do not merge and interpenetrate in the sense of subatomic interpenetration, where particles may be said to share identity intrinsically; only the *appearance* of such interpenetration is created, imposed by an external manipulation. Hence these analogies are no more than mechanical approximations of indivisible wholeness.

At this stage, it might be useful to recall the primary purpose of Bohm's book: to contribute toward the healing (making whole) of our planet by helping overcome fragmentation in all of its forms. With this in mind, it seems we must strengthen our resolve on the matter of wholeness, attempt to go further; mechanical analogies will not suffice. But *can* we develop a nonmechanistic means of representing wholeness? Is there any way to surpass mere analogy, to improve our form of expression by making the representational process harmonize with the content represented? Or must we simply resign ourselves to our inherent limitations in this regard?

Normally, when we use the phrase "it is only an analogy," we are not questioning the underlying analogy of the explicate but are working within it; in such cases, the phrase means that our form of representation is *not explicate* (explicit) *enough*. Improving upon analogy ordinarily involves the achievement of greater precision, a finer articulation of details, the drawing of sharper distinctions so that the model and the modeled may be brought into isomorphic correspondence. Of course, the articulated model itself is to be seen only as a formal device, that is, a set of arbitrary (intrinsically meaningless) symbols manipulated by well-defined rules of operation; the model must not be confused with the substantive reality to which, extrinsically, it is made to refer. In fact, we may say that the differentiative enterprise of modeling depends, in the first instance, on differentiating the model from the modeled or, speaking more generally, the thinking process from the content of thought. To the extent that form can be "disembodied," that is, abstracted from concrete substance, the task of setting limits, demarking boundaries, de-lineating, can proceed. So in the conventional Western sense, to model is to make linear or unfold into clearly distinguishable parts, an operation that is essential for effective analysis. Such a procedure for improving upon analogy is certainly appropriate when the master analogy with which we begin is that of unfoldedness or explicate ordering and, indeed, this has been the program of classical science for the past few hundred years (as discussed above).

Bohm shows us that *today's* circumstances oblige us to begin with a *different* analogy, that of *enfoldedness* or *implicate* ordering. The attempt to improve upon the present analogy solely by further *explication* is clearly an error in judgment that we may attribute to epistemological habit, for now the basic weakness in the form of representation is not that it is not explicate enough but that it is not sufficiently implicate. Just as we rightly have sought to improve on explicate ordering in the past by making it *thoroughly* explicate, it appears we must now strive to make the new order of the implicate more thoroughly *that*. A fuller, more authentic expression of wholeness seems needed. With wholeness as our objective, it is clear that this object of thought cannot be severed from the thinking process that gave rise to it in the manner of conventional modeling, for that would make wholeness into just another "thing out there," a simply autonomous entity extended in space and thus mechanical in nature.

What *would* "improving upon analogy" entail? To reiterate the critical question, *can* any positive steps be taken, or is negation what is called for? Are we to obtain the implicate merely by renouncing the explicate? Then, of course, we could not really say we were improving upon analogical representation, but would have to admit that representation were being abandoned. Perhaps true wholeness can be realized only by giving up all our efforts at representation. Is not representation a necessarily mechanistic endeavor? Would not any form of symbolic expression eventuate in fragmentation? In the first chapter of his book, Bohm provides an historical perspective from which we may begin to ponder these matters.

A Developmental View of Wholeness

Bohm speculates that at the outset, human consciousness was not fragmented but seamlessly whole. In the original state of human awareness, thought and thing were radically fused. Bohm comments that "it is of course impossible to go back to a state of wholeness" (p. 24). Now this I would interpret as an inability to regain our *innocence*. I suggest that the primordial state was an undifferentiated one in which thought and thing were actually *con*fused and that this was followed by a differentiative phase, which was initially beneficial but now has calcified into the destructive parody of itself so well described by Bohm.

But the developmental process being proposed would not stop here. It is both triadic and reflexive in character. Thus, when carried to its proper culmination, a *third* phase would be entered wherein antecedent phases would be *integrated*. On this view, though we cannot simply recapture the inchoate Eden from which we were "cast," we need not be resigned to the contemporary state of fragmentation. We must "go back" to wholeness by going "forward," that is, by *harmonizing* implicate and explicate stages of development.

The concept of triadic orthogenesis (movement from lack of differentiation to differentiation to integration) is reflected in the work of numerous thinkers. It was employed by Herbert Spencer in his interpretation of Darwinian evolution and has been adopted by many students of the growth process, with psychologist Heinz Werner (1948) perhaps enunciating it most explicitly. More recently, consciousness researcher Ken Wilber (1980) helped clarify this principle. He began by noting "the overall sequence of development: from nature to humanity to divinity, from subconscious to selfconscious to superconscious, from prepersonal to personal to transpersonal" (p. 52). In the remainder of his essay, Wilber proceeded to examine the "pre/trans fallacy," a widespread tendency among theorists "to confuse prepersonal [i.e., undifferentiated] and transpersonal [i.e., integrated] dimensions" (p. 53). Finally, the orthogenetic concept being advanced is inherent in psychologist Marie-Louise von Franz's insight into the order of fragmentation she had just criticized: "The rationalism of the 17th century. . . had one advantage after all: it drove father-spirit and mother-matter so far apart that now we can reunite them in a cleaner way" (1975, p. 42).

This "cleaner reunion" is just what is intended by the harmonization of implicate and explicate stages of growth. Instead of merely seeking to surrender ourselves mutely to wholeness (as if we could regain our lost innocence) or attempting a simple explication of it (as if wholeness could be reduced to nonholistic, mechanical expression), we would embark on the third path of *integration*. It follows from the triadic principle set forth that there *should* be a means of improving upon analogy, giving voice to wholeness without robbing it of its intrinsic nature. A nonreductive form of representation should be possible, one that preserves the radically nonlinear, deep-process quality of the indivisibly whole. And the effect of this synthesis of unitive and separative modes of knowing should be deeply therapeutic.

The Dance of Mind

As we shall soon see, Bohm provides a valuable clue on how to achieve the synthesis desired. But I must first confess it is not clear to me that wholeness is promoted with complete consistency throughout the book. A conflicting theme appears to be subtly interwoven. In parts of the book, the impression of an absolute dichotomy is created, an unbridgeable gulf between order of any kind (whether explicate or implicate) and a fundamentally unorderable "meta-wholeness" called the "holomovement." According to this idea, while systems of order and measure should be regarded as originating in the holomovement, the holomovement itself would remain totally implicit, entirely inaccessible to any positive form of knowing whatever. Writing in this vein, Bohm asserts that the holomovement cannot be even approximated or approached. If this contention were taken literally, would we not be led to conclude that implicate order, *as* a form of order, is as far removed from the unorderable, true source of undivided wholeness as explicate order and, therefore, is actually no less mechanistic? Our enthusiasm for the therapeutic promise of the implicate analogy should then be dampened, perhaps giving way to a sense of disillusionment. Indeed—since this underlying theme of the absolute relativity of order clearly would include the *linguistic* ordering constituted by Bohm's book itself—were said theme to be carried to its ultimate conclusion, we would be obliged to devalue the very words and sentences of the book, for these would approach that which is unapproachable no better than if the pages had been left blank!

The basically contradictory character of such relativism is evident: it is a positing that denies all positing and therefore denies *itself*. Thus, after positing the idea that "ultimately, the actual movement of thought embodying any particular notion of totality has to be seen as a process, with ever-changing form and content. . . [a] flux of becoming" (p. 63), Bohm goes on to say that "even this statement about the nature of our thinking is, however, itself only a form in the total process of becoming" (p. 63). Here one is tempted to respond by asking, What about *this* (latter) statement? Shouldn't it too be qualified? In using the notion of absolute process to relativize or qualify the notion of absolute process, the process idea, in effect, is canceled. All that survives is sheer *qualification*, an infinite regress of self-doubt.

But another, far more constructive strain is strongly in evidence in Bohm's work. A relativization of absolute relativism itself can be found. We are trapped in self-negation *only as long as* we cling to the fragmentizing style of knowing in which we seek merely to *explicate* wholeness. The author intimates that this limitation *can* be surmounted, that direct insight into wholeness can be gained by *changing* our style of knowing. "What is required here," says Bohm, "is not an *explanation*. . . [but] an *act of understanding*" (p. 56). Alternatively, he speaks of "intelligent perception," "creative insight," and a "dance of the mind," emphasizing that "such comprehension of the totality. . . is to be considered as an art form, like poetry" (p. 55). Therefore, notwithstanding impressions created elsewhere to the contrary, Bohm does foresee a means of reconciling systems of order and measure with the immeasurable holomovement:

> when the whole field of measure is open to original and creative insight. . . the whole field of measure will come into harmony, as fragmentation within it comes to an end. But original and creative insight within the whole field of measure *is* the action of the immeasurable. . . The measurable and the immeasurable are then in harmony and indeed one sees that they are but different ways of considering the one and undivided whole. (pp. 25–26)

As I view it, Bohm's recommendation is fully in keeping with the path of integration set forth above. To realize wholeness, we can neither maintain the fragmentizing manner of representation prevalent in Western science nor simply become mute but must switch to a mode of expression akin to that found in *art*. If this is the crucial step to be taken in facilitating the healing of our planet, we need to understand more clearly what it involves. To this end, we shall compare conventional scientific and artistic forms of expression in somewhat greater detail.

The Symbol in Science and Art

At the heart of the matter is the usage of symbols. In both science and art, to symbolize is to indicate or refer to so as to objectify or externalize, to make into a distinct content of awareness. As we have seen already, the traditional scientific program of symbolizing (modeling) is predicated on the clean separation (via abstraction) of the symbol from that which is symbolized, of means or process from end product. The intended result is that the content referred to always be strictly explicit, a well-delineated entity extended in space, a sharply fixed object devoid of the flux its union with process would entail. By virtue of the same sheer cleavage, the symbolizing *process* is to be exclusively *im*plicit; the references it makes must never be to itself (except in trivial identity operations) but only to what lies outside of it. Being thus (nontrivially) unreferred to in its own field of operations, the process of symbolizing is solely a formal device, a form without substance.

By contrast, in artistic symbolizing we have "license" to pursue the (re)unification of form and substance, process and content; here we may seek to reconcile the implicate and explicate and so mend the schism. The essential idea is *nontrivial self-reference.* The symbol in art does not merely refer to itself by the trivial identification, $X = X$, and to that which is external to it in a purely arbitrary and mechanical fashion. Rather, the symbol is intrinsically *identified* with its external referent, $X = \text{not-}X$, so that the act of referring beyond itself is at once an act of referring *to* itself. We may say that in all modes of effective artistic expression, the external referent of the symbol is, in some manner, also an *internal* referent indicating the symbol's own inherent form. In this way, the symbolizing process is invested with nontrivial content and content is enlivened by process. It is by thus activating the process dimension that the artist establishes the inner or implicate locus of expression—a mood is created or an emotion conveyed that otherwise would be ineffable. So the implicate is expressed nonreductively in art, made explicit without being divested of its indwelling nature.

The Symbol as an Agent of Healing

The therapeutic consequences of harmonizing inner and outer reality may be seen more plainly and directly in the context of psychoanalysis, where, typically, the central aim is to "make the unconscious conscious." It is true that the psychoanalytic movement developed in a climate of Western rationalism that led it to misconstrue the unconscious, to portray this basically implicate facet of our being in mechanistic terms as an "extended thing" merely to be analyzed. A notable departure from this trend is found in the work of C. G. Jung. In its essence, Jung's program may be viewed as an attempt to systematize the nontrivially self-referential use of symbols for specifically therapeutic purposes.

From the Jungian perspective, the ultimate goal of psychotherapy is to achieve a balance of Self, associated with spirit or *psyche* (that which is implicate), and ego, associated with matter or *physis* (the explicate). A fundamental method recommended for this purpose is that of active imagination (Jung 1969). Deliberately and without just dissolving, the ego becomes receptive to what lies beyond its inner horizon. It does this by allowing aspects of Self to take concrete form as psychically potent, nonreductive images or symbols called "archetypes," representations of the Self that nevertheless are not mere ego contrivances. By thus permitting positive expression to the Self—an uncontrived, intuitively coherent embodiment—the presence of the unconscious is authentically felt within conscious awareness and a therapeutic centering effect is realized.

But to address effectively the all-pervasive problem of fragmentation identified by Bohm, the symbolic realization of *psyche* within *physis* that currently seems required could not be enacted merely within the limited sphere of an individual ego's private way of knowing (as in the conventional practice of Jungian psychotherapy); it would need to be extended to the domain of the collectively shared, *public* order of knowledge, be brought to bear at the foundations of science and mathematics. Von Franz was proposing such an extension in her call for the "cleaner reunion of father-spirit and mother-matter," which she followed up by advising psychotherapists to become students of *physics*, and physicists to adopt a Jungian strategy (1975, p. 45). Consonant with this is physicist Robert Jahn's (1982) suggestion that, in order to attain coherence in contemporary theoretical science, the "analytical, scientific" perspective may need to be integrated with the "creative, aesthetic . . . without sacrificing the integrity of either" (p. 161).

While in traditional depth psychotherapy, active imagination may be beneficially implemented through a model, drawing, or painting individually fashioned by the participant, for the constructive reordering of our *collective* mode of knowing, the most universal form of archetype would appear to be required. Indeed, toward the end of his life, Jung himself apparently saw the need for discovering more universal archetypal forms. And his colleague von Franz carried this work forward after his death, her explorations culminating in the book *Number and Time* (1974). Number is interpreted here in a non-Western, qualitative manner, and hypothesized as the primordial organizing principle for the *unus mundus* or "hidden continuum" from which less-fundamental archetypes emerge.

"Mathematics that is No Longer Mathematics"

David Bohm's own mathematical excursion initially seems motivated less by a search for broadly transformative, integrative symbols than by the desire to differentiate: "to articulate or define [the language of physics]. . .in more detail so that it allows statements of greater precision" (p. 158) regarding the implicate order. Thus Bohm begins with a preliminary effort to describe implicate order algebraically. The enfoldment operation is expressed as a mathematical term that can interact with other terms by multiplication and addition: "If we further introduce a unit operation (one which leaves all operations unaltered in multiplication) and a zero operation (one which leaves all operations unaltered when added), we will have satisfied all the conditions needed for an algebra" (p. 162).

Of course, as Bohm points out, such an algebraic representation would *linearize* implicate order in a manner similar to the algebra already employed in quantum theory. In fact, the notion of simply distinct mathematical operations interacting by addition and multiplication but being *unchanged* by this interaction appears in close accord with Bohm's account of *mechanistic* order: "The world is regarded as constituted of entities which are *outside of each other*, in the sense that they exist independently. . .and interact through forces that do not bring about any changes in their essential natures" (p. 173).

Naturally not content with this explicate approach to the implicate, Bohm proceeds to consider another form of algebraic representation. We are introduced to algebras (like the Clifford system) possessing "nilpotence," that is, non-zero terms that when multiplied become zero!

> One may say that properly nilpotent terms describe movements which ultimately lead to features that vanish. Thus, if we are seeking to describe invariant and relatively permanent features of movement, we should have an algebra that has no properly nilpotent terms. . . if we considered impermanent and non-invariant features (implying algebras with properly nilpotent terms). . .then entirely new orders (not reducible at all to (3 + 1)-dimensional order) may become relevant. (pp. 169–71)

If the necessary conditions for an algebra are satisfied by including operations that assure invariance (e.g., the unit and zero operations), then should we not say that algebras such as the Clifford *transcend* themselves (unless they too are to be dealt with reductively, which would mean reintroducing the invariance assumption at another level)? We build invariance into our formulations because we are intent on distilling a finished product. Our exclusive concern is with a palpable end result, not with the process behind it. We hope only to be able to point "out there," as clearly as we can, to a simply autonomous object (be it physical or mathematical). However, in the notion of nilpotence we find a process that in a literal sense does *not* simply end in a product—the product of multiplication vanishes! May we not say that this process "ends" in *itself*, in a celebration of its own flux? This would mean that it does *not* end, for we only can speak of endings in connection with products.

No doubt the nilpotent term *is* a term or symbol. As such, it possesses a limited or finite aspect; it is not *sheer* process any more than mere product. We have come to realize that while linear systems dichotomize process and product, thoroughly *non*linear ones integrate them. The nilpotent system appears to constitute such an integration. This paradoxical identification of process and product is the essence of the nontrivial self-reference idea discussed above in the context of examining artistic expression. Presently, we may say that in referring to itself, the nilpotent expression is neither trivially reaffirmed as the thing that it is ($X = X$)—for the thing is also a process ($X = $ not-X)—nor has expression been trivially negated (not-X = not-X), since the process is also a "thing" (not-$X = X$), a symbol with positive content and quality.

By so contemplating the nilpotent symbol, we make the transition from number as an empty contrivance having trivial content (a strictly formal means for pointing beyond itself with no meaningful end *in* itself—a quantity without quality) to number as "process quality," as *archetype*. We thus approach the "inner horizon," a mode of comprehension through which the measurable and immeasurable may be brought into harmony. As long as we attempt to describe the implicate in linear (i.e., explicate) terms, seeking to capture it in the field of measure, it will elude us. But by a radically *non*linear, essentially archetypal symbolization of the implicate, "the whole field of measure can be opened to creative insight." In this way we may surpass explication and engage in an *act of understanding* (p. 56). The mind may "dance" and the problem of fragmentation be effectively addressed.

If we were to say that the proper symbolization of the implicate requires "an algebra that is no longer algebra," then, on the *geometric* side of this archetypal mathematics, the required concept would be one in which "space is no longer space, nor time time"—to quote Carl Jung's (1955, p. 90) reflection on the domain of "synchronicity" (acausal connectedness) *[see chapter 8]*. Here we return to the critical issue of continuity and discontinuity. Space—or more generally, *dimension*—is assumed by mechanistic convention to be simply continuous, that is, an infinitely dense thus endlessly divisible, extended regime, whose abutting point elements are completely external to one another. Having considered this long unchallenged notion of dimension in some detail earlier, we recognize it as a major source of fragmentation that rightly cannot be applied at the quantum level, where there is fundamental *dis*continuity indicative of wholeness.

Now the mathematician Charles Muses has questioned the propriety of calling a dimension that is mechanistically conceived a "continuum." In his view, it is illogical to speak of "continuity" when referring to "the possibility of infinite partibility or subdivision . . . [the classical dimension] is not a continuum but actually an infinite *dis*continuum" (1968, p. 37). Making this semantic adjustment, the apparent contradiction of equating quantum discontinuity with wholeness is dispelled. It is evident that, in the quantum case, "discontinuity" does not entail further fragmentation within space, the further subdivision of infinitely divisible dimensionality. Rather, what is "fragmented" is the classical notion of dimensionality itself—the source of fragmentation is fragmented, the *discontinuum* is rendered discontinuous, leaving *in*divisible wholeness!

So it certainly seems we must revamp our way of thinking about dimensionality if we are to begin to comprehend the quantum reality. Muses (1975b) has contributed to this goal by introducing the important idea of *fractional dimensions*, dimensions with *less than* infinite density *[see chapter 5]*. In Muses's algebraic formulation of fractional dimensionality, real number systems are superseded by what he terms "hypernumbers." This is of considerable interest to us because the hypernumber idea appears to include the notion of nilpotence discussed by Bohm.

Bohm himself, in ending his book with a call for a *multidimensional* portrayal of the quantum physical connectedness, does not directly question the standard topological notion of integer dimensionality. Yet a hint of fractional dimensionality is provided in his comment that "the idea of space as constituted of a set of unique and well-defined points, related topologically by a set of neighborhoods and metrically by a definition of distance, is no longer adequate" (p. 166). The concept of fractional dimension also may be inferred from Bohm's suggestion that whereas implicate order cannot usefully be described by continuous curves, it may be approximated better by *Brownian* curves (geometric models of Brownian motion; see Bohm 1980, pp. 127–28). In mathematician Benoit Mandelbrot's book on *fractals* (1977), the Brownian curve is offered as the prime illustration of nonclassical, nontopological, "fractal dimensionality."[3]

Taking our cue from the explorations of David Bohm, we are led to the farthest frontier of our fragmented way of knowing. Here a glimpse is obtained of a "mathematics that is no longer mathematics" or, in the broadest terms, of a manner of representation in which the symbol no longer serves as an agent of fragmentation but plays an essentially *archetypal* role, is a catalyst of wholeness promising in its universality, the therapeutic transformation of the human community. Therefore, it is my conviction that at this critical moment in human history, we would do well to take our cue from Bohm.

13

The Bohm/Rosen Correspondence (1983)

July 25, 1983

Dear Professor Bohm,

I greatly appreciated. . .your letter of response to me. Enclosed you will find a transcript of your letter [ca. June 15, 1983] with my interpolated comments appearing in brackets; I offer these in a spirit of exploration, claiming no solid foothold in this difficult "terrain". . .

Sincere regards,
Steve Rosen

[ca. June 15, 1983]

Dear Dr. Rosen,

Thanks very much for your letter, and your articles discussing the implicate order, which I have read with great interest. I am enclosing part of a talk I gave on the implicate order, which I hope that you can read. It contains the new notion of *soma-significance* (or somato-significance), which I think is relevant to the questions under discussion. This notion implies that the somatic (or the physical) and its significance (which is mental) are two aspects of one reality. In your terms, each aspect reveals (or displays, i.e. unfolds) the whole of reality. Each becomes the other (i.e., the content of each passes into that of the other). Each reflects (or reveals or shows) the other. Each *is* the other, insofar as each is an inseparable aspect of the same ground of reality as a whole. The two aspects are one, and one with reality (though they are also distinguished).

[[Yes, the above is clear to me and I wholly concur.]]

I think that the notion of *active meaning* is a proper extension of that of *active information*. Information is in "bits," represented by Boolean algebra. But meaning is a much vaster notion, which encompasses information as a small part of itself. The algebra of meaning would be at the very least, an infinite

223

matrix algebra. For example, a projection operator, $P^2 = P$, when diagonalized, can represent a Boolian algebra, with its "yes-no questions" representing "bits" of information. But the whole algebra describes movements, projections, reflections and transformations, which give the *meanings* of these "bits" of information. Such algebras include quantum theory and much that is beyond. So I am proposing:

Reality *is* a totality of active meaning. Our own meanings are only a part of this. What the world *means* to me determines my intentions, and therefore, my actions toward it. What I mean to the world determines its actions toward me. This holds not only for other people, but also for inanimate matter.

To be somatic or physical is a particular kind of meaning. This is distinguished from *subtle* meaning. The root of "subtle" is "sub-texere," meaning "woven from underneath" or "finely woven." Its meaning is "rarefied, delicate, highly refined, elusive, undefinable, intangible." My proposal is that reality has two aspects, gross and subtle. Of course, this is relative, since what is subtle on one level is gross when considered on a more subtle level. [[Could this relativism be viewed from a developmental or evolutionary perspective, namely, that our development entails a "somatizing" or embodying of that which was ineffably subtle?]] Soma is essentially the grosser aspect of reality, and significance, its subtler aspect. My proposal is:

Ultimately, the gross is embedded or encompassed in the subtle (soma in significance or the physical in something mind-like, such as universal mind). So reality is apprehended through two distinct aspects, subtle and gross, but ultimately, the subtle encompasses the gross.

I hope that the part of my talk enclosed with this letter will further explain soma-significance. I suggest that you read it now before going on with the rest of the letter.

Now to comment on a few points you have raised. You ask for a symbol which does not separate observer and observed (one whose *significance* is therefore directly in the implicate order, rather than in an explicate order, whose content is an analogy with the implicate order). You bring in the Moebius strip here. I see what you *mean* by the Moebius strip, which is an explicate model of ultimate identity of two distinct sides, such as explicate and implicate. But it still suffers from being an explicate model [[agreed; perhaps its higher order counterpart, the Klein bottle, when not treated reductively, as it is in conventional topology, but considered "poetically" (i.e., inwardly, rather than merely as an object to be delineated), may have a more directly implicate significance]] as well as because the two sides are too much alike [[actually, I believe the sides of the Moebius strip are more different than, say, the two sides of an ordinary, untwisted ring—the former are related to each other enantiomorphically, i.e., as left and right hand, rather than as simple identities; i.e. they are both the same *and* they are different]]. You try to remedy this by considering Escher's drawing of the picture in a museum, which opens up into a world that ultimately contains the museum and the one who is viewing the picture *[see fig. 3.1]*. This is better because it differs in an important way from the Moebius strip. For in it, the oneness of viewer and what is viewed is implied by going in an *unbroken* way [[I've added emphasis to the unbrokenness you observe since I believe it is a critical point, as later comments will reflect]] to a more encompassing level, which contains the starting point as a special, distinguished element. Moreover, the new level is more subtle, since it implies a single mind or consciousness, in which observer and observed are explicate contents of the picture as a whole, while this mind is actually "in" a body, which is also "observing" the picture with its content of the relationship of the "museum observer" in the "museum world." Moreover, the actual mind and actual body are related in a way similar to that in which the "museum observer" and "museum world" are related. In a way, the Escher drawing functions as a *metaphor*.

A metaphor is an *implied comparison*. It is to be distinguished from a *simile*, which is an explicit comparison. In a simile, one says A is similar to B in certain explicit respects, different in others. The elaboration of a simile ultimately gives rise to *prose*, which is an *indefinitely extended simile*, i.e. to talk prose is to say (or imply) that the world is *like* what I say (though also *unlike*). In a metaphor, it is essential that the similarity is left implicated or unstated. I suggest that *the meaning of a metaphor is therefore directly apprehended in an implicate order of the mind.* So in poetry, it is possible to engage the implicate "dance of the mind."

To go further, I suggest that *creative mathematics* is a kind of metaphor. To see this, consider the equation, A = B. Now, to equate two things that are identical (in the sense of not being different) (A = A) is a triviality. The only value of an equation is to equate two things that are different. If we look at this *prosaically*, we can say that A = B in *certain respects* (but not in others). We then treat the equation as a simile. But in the act of creative perception of a *new meaning* in mathematics, we are poetic and not prosaic. For example, consider Newtonian dynamics of particles and the Hamilton-Jacobi theory of waves. These are very different in concept. When we equate them, the meaning is, in the first instance, sensed in the implicate order of the mind as a deeply moving perception. It has vast, as yet undefined meaning, which can include relativity, quantum theory, and much that is beyond both. Gradually, some of this is made explicit, and the poetic vision crystallizes and "cools down" into "prose" (i.e., the ordinary ideas of accepted physics).

[[Is the initial insight into the unity of opposites (e.g., particle and wave) really mathematical as such, or is the doing of mathematics the process whereby a *pre*-mathematical insight is "cooled down"? I suggest that the mathematical enterprise ordinarily serves the function of turning poetry into prose, but that a "poetic mathematics" *should* be possible, and that it would depend on *preserving* the poetic quality, keeping the "poetry" in the elaboration of the initial creative insight, not allowing the poetic vision completely to "cool down."

[[You have observed in effect, that *all* forms of assertion *originate* in the implicate and therefore, are expressions of the implicate. Consequently, there is no denying that prosaic statements *are* expressions of the implicate. The point is that *poetic* statements are unique in that they express the implicate *directly and immediately* by consciously (explicitly) embodying ("bodying forth") the implicate in their own inherent form. By such embodiment (which would need to be intuitively unitary, a coherent, unbroken expression), the implicate is kept in the foreground, does not recede from consciousness, is not permitted to "cool down."

[[What of the equation A = B, when not intended in the merely prosaic sense of indicating a certain restricted equivalence? Perhaps writing the equation A = B might be thought of more as an act of writing *about* poetry than as a poetic action in itself. I propose that in an actual poetic statement of the identity of opposites, the natural (intuitive) basis for the act of identifying A and B would be displayed, explicitly embodied, consciously symbolized or expressed. I suggest that when this is not done, the *content* of the statement may be properly poetic but the form would not do justice to this content; it would constitute an attempt to refer to a poetic relation in a discursive, non-poetic language system. To say "A = B" or "I am you" (and to mean it in more than just the prosaic sense of equivalence in *certain respects*) may pave the way for a positive poetic statement, but perhaps such statements—because they are being made in the *form* of the old, prosaic language—are not poetic fulfillments in themselves but do more by way of simply revealing the *limitations* of prosaic form. In other words, I would distinguish prosaic expressions that in their paradoxical (prose-denying) way may *suggest* poetry, from positive poetic expressions, those in which a poetic *form* is employed that does justice to the poetic content. I would propose that the Klein bottle relation (if not reduced to the prose of classical topology) would constitute a poetic expression because it does not simply *posit* the identity of opposites but makes explicit, expresses in the mathematical *form* itself, the intuitive basis for the identification of opposites. In this way, form and content are unified, such unification being the sine qua non of authentic poetic expression. This line of approach may provide a basis for distinguishing positively poetic mathematics from a prosaic mathematics that by self-negation, may only suggest poetry.]]

I want to suggest that creativity is quite generally a perception of new meanings that encompass older ones, showing their limits, and going beyond. Since meaning is *active* (as explained in my talk), the perception of a new meaning is the creation of a new active factor in reality. *So to see a new meaning is actively to change the world.* [[Agreed.]] The world is not a *complete* whole, which is there, waiting to have its meaning perceived by us. Rather, reality (animate and inanimate) *is* active meaning. This is never complete. [[Here, our previous dialogue on the nature of infinity comes to mind; the approach I've called the "actualiz*ing* infinity," gives rise I believe, to a notion neither of simple completeness (which would be associated with an already actualized infinity) nor incompleteness (an entirely open, unactualizable infinity) but of *relative* incompleteness.]] In natural processes, a radically new species is, in essence, new active meaning in the world (perhaps perceived in a universal mind). Our own intelligence can contribute to reality in this way also.

The challenge before us (requiring a creative perception of new meaning) is to use language and thought in such a way that we do not, at the very outset, tacitly take for granted the notion that the world exists "in itself" apart from its meaning (to itself, to us, to others, etc. etc.). Nor should we suppose that meaning exists in a purely "spiritual" sphere, apart from the reality that is *meant*. Rather, we have to encompass the notion of *soma-significance* deeply and pervasively in our active mental operation, so deeply and pervasively that it is included even in our statements about physical theory, . . . , meta mathematics, etc. etc. [[I would agree fully with all of this but would emphasize the primary importance of the essentially artistic question of the *form* through which soma-significance is expressed. I believe that beyond the mere paradoxical use of our prosaic language, an actively poetic form of expression is needed (as proposed above); in my "neo-intuitionist" approach (offered as an alternative to formalism, the latter perhaps being seen as an implicit recognition of the need for poetry that nevertheless cannot break out of prose and so ends in *non*-intuitive, arbitrary axiomatics), the intuition of soma-significance would be brought out positively, made poetically explicit so its meaning can be directly apprehended.]]

What I am writing now, at this very moment, is a form of meaning. But at the same time, and in this same act, it is physical. Thus, it consists of patterns of ink on paper, made by muscular movements, initiated by nervous impulses, organized in a brain affected chemically by emotions and in other ways by perceptions of reality, external and internal, and by relationships with other people (including you who are the intended reader). This physical (somatic) side of thought is not ordinarily present to awareness, at least to any significant extent. It takes place mainly as what Polanyi called *tacit knowing* (e.g. bicycle riding). This tacit knowledge is thought on the side of its *concreteness*. Such concreteness includes not only a somatic aspect of implicate order, but also a vast set of tacit *meanings*, which are also implicate. It is all this which gives thought its power to engage with physical reality, and to go beyond the given into new ranges of meanings. And all this is *one* with all other aspects of thought, especially its symbolic activity and its explicate content.

Now, is it possible to have, as you implicitly call for, a symbol that makes us instantly aware of all this, in its concrete actuality, moment by moment? If not, then the explicate content will, as usually happens, be divorced from the vast implicate activity of soma-significance. What is needed is a kind of *proprioceptive thought*. As is well known, the nervous system is so "wired up" that, as impulses to move originate in the brain and spread through the body, the system is aware of the distinction between its own activities and movements originating independently. This is essential to maintain order (e.g. in dizziness or vertigo, this distinction breaks down, and proper movement becomes almost impossible). But there is no "wiring" of nerves to tell the brain of all those activities which originate in the action of thinking.

This brings about a great deal of confusion. In particular, when the *consequences* of thinking are "picked up" and recognized, they are falsely attributed to a source independent of thought. If they are in the body, they will be attributed to an entity [called] "me." Since these consequences (e.g. chemical, hormonal, etc.) are sensed as affecting the course of thought, this leads to the further mistake of attributing the activity of thinking to this entity, the "me," who is now supposed to be the source of thought (whereas the truth is that "the me" and thought are one and inseparable). In this way of thinking, not only "the me," but even more, "the world" have to be projected as separate from thought about them. If a person does not perceive the oneness of "the me" with thought, it will be futile to try to think in terms of the oneness of thought about the world and the world itself. But the whole of society (including the community of scientists, mathematicians, etc.) perceives "the me" as a reality independent of thought. So it seems that the way to do what you want to do is blocked. Something more is needed than a change of *symbolic structures* to change a deep structure like this in a fundamental way. [[Yet note that I am not really calling for a change from one (explicate) symbolic structure to another, but a change in our fundamental *way of symbolizing* from one that is merely prosaic to one that is genuinely poetic and therefore should be able to help mend the schism between "the me" and the thinking process.]]

Something beyond thought is needed to break through this impasse. Krishnamurti (whom I know quite well) has gone into this question very deeply. What is beyond thought includes awareness, attention, intelligence, insight. (We take intelligence here as a perceptive act, a creative perception of deep inward meaning.) [[Could we view the creative perception of intelligence and the deepest forms of poetic symbolizing as interpenetrating aspects of each other?]]

Even in dealing with ordinary material objects, thought and its habitual responses as tacit knowledge are not enough. We need an *awareness* of this object. Krishnamurti indicates that this awareness is a function of the learner that is not conditioned; i.e. it is constantly open to what is new and different from the known. Awareness modifies the content of thought, in accordance with the actual fact. The suggestion is that attention goes further, and that it is already a link between "the brain" (i.e., the somatic) and what is "beyond the brain" (i.e., a meaning not restricted to a base, as recorded in matter). Attention "stretches toward" the object. Here there is an analogy to what is done with a physical object. When we are actively aware of it and attend to it, we may also be touching it, probing it physically and optically, thus *learning about it. The suggestion is that awareness and attention can probe the implicate somatic movements of brain and nervous system in a similar way.* The idea is simply this:

Thought has to be (at least potentially) in a somatic contact with its real object, if it is not to degenerate into fantasy, illusion or worse. The nature of this contact is clear for external objects. It is less clear for what is going on inside the body, still less clear for thought itself. But since thought has a somatic side, *it is in principle possible for thought to be aware of its own somatic movements, and thus to correct its present defect of not being proprioceptive.* Without such awareness, thought will form illusory notions about itself, based largely on imagination. And this will lead to confusion in its activity.

Is it really possible to be aware of the implicate order of activity of the nervous system? There is much evidence that this is possible. For example, in an experiment reported in *The New Scientist*, a fine wire was attached to a single nerve in the finger, and this was attached through an amplifier to a speaker. Every time the nerve worked, the speaker clicked. The person was then able in a way that he could not describe (i.e., subtle, elusive, undefinable) to make the speaker click. Eventually, he could play a tune on it.

Of course, this is a kind of biofeedback. But I prefer the word *display*, whose root, dis-plicare, means "to unfold." But the unfolding is intended to *reveal* what is enfolded, and is not done for its own sake. Whenever the implicate order of operation of the nervous system is *displayed*, the nervous system is aware of itself and can act on itself and so, proprioception is possible.

A mirror image of me is a display of me. It enables me to be aware of my body and its movements. Thus, I do not attempt to shave the image, but rather, I use the image to guide me in shaving *myself*.

A similar activity is possible inwardly. Usually, the contents of consciousness (images, feelings, ideas, etc.) are unfolded, because they are regarded as significant in their own right (generally, because they reflect or indicate some object beyond these contents). But now, I can regard these contents as a display of *me*. However, this display is not a direct correspondence of an explicate image with an explicate "me" (who would be somewhere inside the body). Unlike what happens with the mirror, it is displaying an implicate order (as the loudspeaker does, when attached to a nerve). When the meaning of the displayed is sensed properly in this way, the mind can (as with the speaker) direct attention to the implicate order of somatic activity out of which the contents of consciousness arise. And so consciousness (which includes thought, feeling, desire, will, impulse to act, etc.) can be proprioceptive.

In British television, a test card is often transmitted. The images on this card are not significant in themselves. Rather, they can display errors in the functioning of the electronic equipment, and thus direct the somatic activities needed to bring this equipment to order. Similarly, awareness of and attention to the content of consciousness can direct the nervous system to bring its own activity to order (according to Krishnamurti, this is what is essential for the beginning of true meditation). When there is such order, thought can be part of the action of intelligence. Otherwise, thought is generally dominated by the response of memory, which acts as a sort of computer programme. When there is no proprioception, the outgoing activity of the programme is mistaken for incoming perception, and thus, a self-enclosed feedback loop is set up, which is very difficult to end. This feedback prevents the "reception" of the action of intelligence by the brain and nervous system (rather as a radio receiver caught in feedback drowns out the signal with its own self generated noise).

Of course, the feedback is only a predisposition or tendency. And so, there may fortuitously be moments of clarity, when it does not operate. But sooner or later, the noise will start up again, equally fortuitously. And then, the action of intelligence in thought will be swept into the destructive confusion of the feedback loop. For clearly, memory is not capable of awareness and attention. And so, its feedback loop is a mere superficial imitation of the true perceptive feedback, that comes with awareness of and attention to the display of consciousness, seen as revealing its enfolded activity.

To sum up, then, consciousness has to be aware of its own implicate activity, in which its content originates. Only then can thought be counted on to be the action of intelligence. Otherwise, it is subject to fortuitous distortion, by its own feedback loops. Meditation, beginning in awareness of and attention to the whole *activity* of the content is the beginning of the required new mode of consciousness, that is free of memory-based feedback loops (which are evidently a kind of *conditioning* that is very destructive).

[[I have spoken of "properly poetic symbolizing," by which I have meant a form of symbolizing that is non-trivially, meaningfully self-referential (this was a key point in my essay on your book *[chapter 12]*). The content of such symbolizing does not simply reflect or indicate some object that is beyond it, but directs attention back to itself, directs attention *inwardly* to its implicate source. A symbolic action that merely points beyond itself effects a severance of that which is pointed to or thought from the thinking process. On the other hand, a symbolic action that points non-trivially to itself, concretizes or embodies its referent, i.e., it displays its explicate referent within the implicate context out of which it (the referent) unfolds.

[[I suggest that such self-referential symbolizing may be viewed as an inseparable aspect of the proprioceptive, meditative process you describe (perhaps such symbolizing can be seen as the "somatic" aspect of a soma-significance relationship whose "significance" is the act of meditation, with the understanding that poetic symbolizing and meditative being are one and inseparable, though at the same time different). Can the content to be concretized or embodied be the seamless totality itself? The position could be taken that in this case, no form of symbolic knowing could suffice but that we might be able to transcend knowing entirely and grasp totality in an act of pure meditation, as it were. But I would propose (for reasons to be developed below) that if a genuine meditation on totality could be achieved, like lesser forms of meditation, it too would have a (poetically) symbolic aspect, an aspect that would disclose authentically the "contour" of the totality.]]

Evidently, an orderly "dance of the mind" depends on the action of intelligence (otherwise it will resemble the "dance of a drunken man"). This order comes from the depths of the implicate order, beyond thought. We can present a rational description of what is involved in it (as I hope that I have been doing in this letter). Such a description helps to neutralize some of the false and irrational ideas on the subject that are current in society. And thus, it may help bring about a state of mind in which reason *does not attempt to oppose the action of intelligence.* But something beyond reason is needed for this, an *actual awareness* of the contents of consciousness as displaying its inner (implicate) activity, along with an *actual attention* that "stretches toward" this activity in an indescribably subtle way.

In your article *[chapter 12]*, you question my attempts to say that ultimately, thought cannot grasp totality [[actually, I have strongly agreed with you that no act of merely prosaic thought can grasp the totality; it was only when I felt that the impression was being given that there was *no way whatever* to grasp totality, that I raised my question]], stating that this means ultimately the collapse of all mental activity, including what we are doing with the implicate order. My answer is that thought always requires what is beyond thought (i.e. awareness, attention, etc.) for its orderly action [[and I would agree with this, provided that the distinction be made between prosaic and poetic thought (in the sense of poetry discussed above), and that poetic symbolizing be viewed as meditative in nature, the "somatic" aspect of the meditative act]]. What we are doing with the implicate order will make full sense only if it is done with *proprioceptive thought.* Otherwise, thought about the implicate order can be as confused as thought about anything else, when it lacks proprioception. If there *is* such proprioception, then our discussion of the implicate order may be valid, without implying that thought can capture what is absolute or total.

[[Granting that totality cannot be *captured* (as a mere thing might be captured, enclosed, reduced to simple finitude), is the implication here that even proprioceptive thought could not approach a *grasp* of totality? It seems to me that proprioception could not confer the order and validity it does without being at least in incipient contact with totality.

[[If it is assumed that a grasp of totality *can* be approached by going beyond ordinary (prosaic) symbolic thought, it could be helpful to clarify the question of the *direction* to be taken in this "movement." One might think simply of going *back*, returning to primal wholeness, the state of pure being existing prior to the emergence of symbolic knowing. But as you wrote in your book, "it is of course impossible to go back to a state of wholeness" *[Bohm 1980, p. 24]*. The suggestion I gave in my essay is that we might "go back" to wholeness by going "forward." In the new essay enclosed, I further elaborate on what such a "movement" would entail. The idea in essence is that instead of symbolic operations merely dissolving in the operation of a sheer intelligence (which would constitute a simple going back), the two would come to be actively understood as different aspects of one and the same operation: an *integration* of symbolic knowing and intelligent being would be realized at the deepest level. Here, the symbol would no longer be merely a symbol but, through radically poetic proprioception, would metamorphose, be suffused with the aspect of the infinite ground; by the same token, the ground—while of course not becoming merely a fixed form that could be captured prosaically, reduced to finite symbolic description—would be no inchoate Eden either, no sheer, formless, pristine implicateness, but would take on a certain contour or "shape." I am suggesting that at this "face of integration," the boundary between knowing and being is at once no boundary at all, so we have neither knowing nor being alone, nor some simple combination of them, but a *synthesis* that is qualitatively more than both.]]

I think it is essential here that thought is limited by ignorance. No matter how far it goes, there is still the unknown, what it does not cover. In certain areas, the unknown has negligible consequences, and so, such ignorance may itself be left out of account, i.e., ignored. But as the domain is extended, one sooner or later meets the unknown.

This is evident in the history of natural science. It is also evident that the symbolic content of thought does not contain *concrete* knowledge of the implicate order of somatic activity in which it arises. And according to Gödel's theorem, one may infer that this content always depends on a vast range of unknown presuppositions, which have tacitly been taken for granted. Finally, it depends on inward depths of meaning in the implicate order that are too subtle to be explicated to any significant extent [[yes, not explained but approached perhaps, through an act of understanding which I suggest would entail a poetically symbolic aspect]]. So, it does not seem reasonable to suppose that the symbolic structure of thought can capture the essential structure of its own "dance," in which the symbols are unfolded. Here, one needs awareness, attention, the action of intelligence-insight, etc.

So it seems to me that it is reasonable to propose, as I have done here, that thought does not capture the absolute. Of course, at the level of content, it is *only* a proposal. What this proposal calls for is that each person test it *for himself* through awareness, attention, etc. My suggestion is that *proprioceptive thought is directly and immediately aware of its own limited nature.* Thus, as I might make all sorts of proposals about bicycle riding, for example, these would, in themselves be mere proposals, until they were tested in actual bicycle riding. Similarly, proposals about the nature of thought have to be tested in awareness of and attention to its actual "dance" in the implicate order. This holds for what I said about direct perception of the limited nature of thought.

While I therefore think that what can be done with symbols is limited, I nevertheless agree with you that we could usefully explore the question of whether we can find *better* symbols to indicate the state of affairs described above. Here, I would like to consider the videogame, as described in my Syracuse lectures. It constitutes two levels of implicate order (that are close to those appearing in modern quantum mechanical field theory). The screen is the first implicate order, and the computer programme is the second implicate order. [[As I wrote in my essay on your book, analogies certainly can be useful ways of indirectly indicating implicate order and it would be helpful to develop the most effective analogies we can. But my suggestion is that there is a deeper sense in which we can improve our symbolizing which involves, not just inventing better ways of analogically (indirectly) indicating implicate order, but entails discovering a way to indicate it more directly through deeply poetic ("archetypal") expression.]]

The second implicate order organizes the first, as I show below:

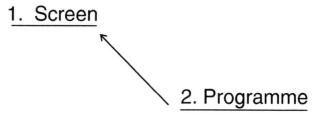

But now, there is a third implicate order, the consciousness of the human being, who is playing the game (and who could also change the programme). He responds to the *display* on the screen by pushing buttons or otherwise affecting the programme and its activity. I indicate this below:

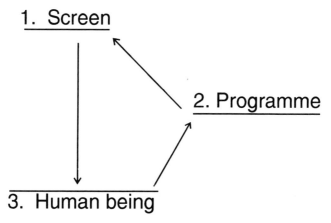

Here, a cycle is set up, and so, there can be the dynamics of a *game*.

I suggest that this is similar in some ways to Escher's drawing, especially if we view it in a direction *opposite* to that of the arrows indicating the order of action. The limited content of the screen expands to a much greater and more subtle content of a programme, and this expands again into the still greater and more subtle content of human consciousness, which in turn has both the screen and the programme as part of this content. The general idea is that higher implicate orders have lower orders as part of their content, and this allows for the "Moebius strip" characteristics that you talk about.

[[In his writing about his work, M. C. Escher emphasized the importance of "unity of expression" (I discuss this matter in my paper on "meta-modeling" [Rosen 1981; see also chapter 11]). To restate my proposition in this light: the question of whether the symbolic expression is or is not poetic, or at a deeper level, is or is not archetypal, depends on whether there is "unity of expression" in the symbolizing, i.e. whether the content referred to—in this case, levels of implication—is positively embodied in the very form that does the referring (which would set up a kind of "inner resonance" making the content more than a mere externality), or whether the content or product displayed is unrelated, or only arbitrarily related to the displaying process, in which case, whatever this content, it would be merely an unfolded product severed from the process which gave rise to it; here I'm suggesting that the *style* or *form* of the symbolic expression is critically important. If we are attempting to express an implicate relation, one entailing the notion of mutual interpenetration, an identity of opposites, then I believe a *form* of expression would have to be chosen that would positively disclose this identity within itself, creating an unbroken flow from form to content so as to unify these. It is for this reason that I have found a naturally self-penetrating form like the Klein bottle to be of special value, a "properly poetic" way of referring to the implicate: it refers to the implicate by referring to itself. Again please note that I'm not suggesting that symbolic expression *by itself* can give direct access to the implicate; the idea is that properly poetic symbolism would operate as the content *aspect* of a properly proprioceptive state of awareness. In this regard, I find it interesting that physical models of the Klein bottle are *incomplete*, apparently because the self-penetration cannot be smoothly (unbrokenly) executed without recourse to an "added dimension of space" (compare this with the incompleteness of the Escher print—the circular blank area in the center [fig. 3.1]. Would it be too fanciful to suppose that the actual "completion" of the Klein bottle would require precisely the mode of meditative awareness, attention, and intelligence of which you speak, that the "hole" in the Klein bottle is a kind of invitation to fuse prajna (identified by D. T. Suzuki as the epistemological, and therefore form-bearing aspect of Zen "no-mind") with dhyana (meditation)? [Prajna *and* dhyana *are discussed in chapter 11.*]]]

I would call the above a *basic soma-significant loop*. It would probably be approximated with a third computer, as below:

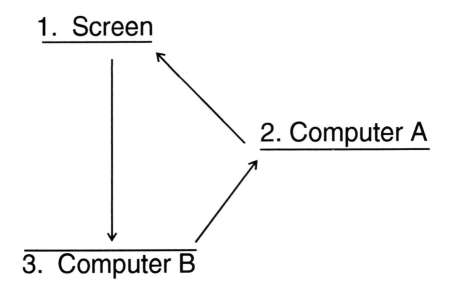

1. Screen

2. Computer A

3. Computer B

Or more generally:

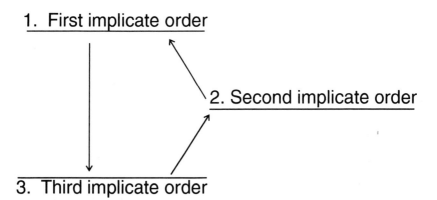

1. First implicate order

2. Second implicate order

3. Third implicate order

To go further we need a fourth implicate order, whose partial content includes the first three, and so on indefinitely. [[Here is the important issue that has come up for us in the past about how one represents the infinite regress. It relates to J. W. Dunne's problem of the painter painting the picture of the painter, and my proposed "Moebius/Klein" (and Escher "Print Gallery") solution—again, it's a question, not of what is represented, but of *how*. From a formalist's point of view, "how" might not critically matter, but from my "neo-intuitionist's" standpoint, the form of symbolic expression is raised to central significance—the choice is not arbitrary. And I propose that were an appropriate, "properly poetic" choice to be made, infinity no longer would be an endless regress (as it must be in prosaic symbolizing), no longer always beyond one's reach; instead, it would become an "actualiz*ing*" infinity.]] However, we propose that in a certain limited domain, the proper implicate orders may be ignored. More generally, the form of experience is such as to make possible the ignoring of the infinity in certain limited contexts. So, each use of thought is, at each moment, possibly a proposal, to be tested, from moment to moment. Some proposals, when tested, are treated as well founded suppositions, not needing more testing, until some evidence or contradiction is obtained. *Even what is said here is such a proposal.* I assert that this proposal is *consistent* as far as we have now tested, but that it could in principle require change or radical transformation.

I do not in this way deny the absolute or the totality. Rather, I propose that this may be contacted in the action of the mind beyond thought. This proposal can also be tested, in ways that have been described here. Also, I do not deny that symbolism may be extended indefinitely. But I propose that however far this goes, such extensions are limited, and do not grasp the actual totality. They may function as metaphors, or else as displays, which guide attention and awareness to those factors that *prevent* the grasp of totality (i.e., the automatic function of thought as response of memory in a feedback loop). So there is always room, I propose, for an extension of our symbolism.

This letter has become rather long, so I'll end it now. I would appreciate your comments.

Yours,
David Bohm

[[To summarize my thoughts and responses (not too long-windedly or redundantly, I hope):

There are three general propositions I would offer for your consideration:

1. that proprioceptive thought does have a symbolic aspect;
2. that through such proprioception we may approach a direct grasping of totality;
3. that in so grasping totality, the relativistic notion of an entirely open, unapproachable infinity, an infinite regress forever stretching beyond our finite reach, is itself relativized (this is the "second principle of the relativity of the infinite" discussed in my paper on "The Concept of the Infinite" [*chapter 4*]).

[[Accordingly, I see certain poetic symbols, deeply self-referential, archetypal expressions as playing a more primary role than just guiding attention and awareness to factors that prevent the grasp of totality. I see them as directly engaging totality, *if* they are operating in conjunction with, as an aspect of, the proper meditative state. Again, I would propose that such symbolizing, and concrete, meditative being, may become one at an ultimate level, may merge and interpenetrate as aspects of a single reality that is the totality (in relation to human possibilities, human evolution), with neither aspect dissolving in the other (as, in your description, soma and significance interpenetrate without dissolving). In the occurrence of such a merger, knowing and being would transform each other: the symbol no longer would be a *mere* symbol but would poetically surpass itself, and pristine being (the primal wholeness to which, as such, we cannot go back) would take on a symbolic aspect. The point I would emphasize is that this ultimate interpenetration would give us neither knowing nor being as commonly understood, nor a mere admixture of these, but a *synthesis* that would be qualitatively more than both. Perhaps we could say that the totality to be grasped is non-symbolic, concrete, grounded being, and *at the same time*, symbolic, abstract, transcendent knowing (it is both immanent, eternally within, there to be discovered, yet also, emergent, novel, an outcome of the creative advance of human consciousness). I'm proposing then, that the status of knowing be elevated to "equal partnership" with being, that totality is realized, not by leaving knowing behind for being once the former has guided attention and awareness, but by working *through* knowing down to its deepest roots, out to its farthest frontiers, where we may discover/ create its old/new entwinement with being.

[[I further feel that the planetary healing now so badly needed might occur if enough of us could so realize totality. I would suggest that such a healing would entail a step in evolution to a new phylogenetic level. Here we would find, not man supremely wise, but "superman" supremely *ignorant*—a whole new vista of the unknown would open to the self-created, novel species. But if such an infinite vista is to open at a new phylogenetic level, I believe we must achieve a *closure* on the infinite at *this* level, actualize our human totality. In general terms, instead of a simply closed or simply open infinity, i.e., a simple completeness ("superman supremely, perpetually wise") or incompleteness ("man forever ignorant"), I am proposing a "*relative* incompleteness" (= "actualizing infinity"). I agree that ultimately, there is *always* an opening up, a new flowering, but am suggesting that phases of "closure" or consolidation always must be interpolated, that such consolidation is a necessary requirement for fresh unfoldment, that we cannot go from "flower to flower" without enfolding a "seed." The proposal of "relative incompleteness" is directed at overcoming the underlying dualism of the complete and the incomplete.]]

AUGUST-SEPTEMBER

September 26, 1983
Dear Professor Bohm,
    I appreciated receiving your two part *[August]* letter (a transcript of which is enclosed with my interspersed comments) . . .
                                Sincerely,
                                Steve Rosen

[ca. August 15, 1983]
Dear Dr. Rosen,
    Thank you very much for your *[July]* letter, . . . plus your transcript of my previous *[June]* letter, interspersed with your comments.
    Your attempt to use the computer as a "poetic" symbol of oneness, which includes self reference, is very interesting. [[Actually, I see the computer more as a potential *facilitator* of poetic symbolizing than as a poetic symbol in itself.]] However, I am afraid that I am a bit prejudiced against computers in general. I regard them as the ultimate of the prosaic. [[I agree that

by themselves, they are entirely prosaic.]] The reason is that they are based on the concept of *information*, with no explicit attention to *meaning*. But information divorced from meaning is about as prosaic a theory as it is possible to have.

Actually, I think that an extension of the information concept is possible, and is indeed suggested by the quantum theory. Rather, let us say that a new concept of *meaning* is possible, within which information is contained as a natural subdivision.

Let me begin with a projection operator, $P^2 = P$. It has eigenvalues, 0 and/or 1. Thus, the eigenvalues can represent information ("yes" or "no" to certain questions, "stop" or "go" to certain activities, etc.). But as a matrix operator, it may have a tremendous number (actually infinite) of *indices*. These I shall denote collectively by α. Thus we write

$$P = P_{\alpha\beta}$$

I now suggest that $P_{\alpha\beta}$ may be interpreted as the *meaning* of the "bits" of information, represented by the eigenvalues.

This notion can actually be applied even to present computers. (Because computer scientists are now dazzled by the concept of information, they hardly ever think of meaning at all.) However, to make things simple, let us consider a matrix, whose eigenvalues are all zero, except for a single one. We denote the normalized eigenvector corresponding to that eigenvalue as V_{α}. We can then write:

$$P_{\alpha\beta} = V_{\alpha}V_{\beta}$$

(assuming the vector to be real).

Let me now consider an element, E, of a computer, that is either "on" or "off."

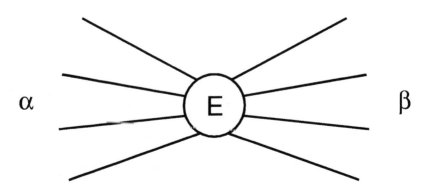

α E β

Let it *[E]* be connected to a variety of inputs, α (shown above on the left) and a variety of outputs, β (shown on the right). We can represent the relationship of input to out by:

$$M_{\alpha\beta} = V'_\alpha E_{ij} W'_\beta.$$

If $E_{ij} = 1$, the input V'_α will go to W'_β and if $E_{ij} = 0$, nothing happens. So the computer element *transforms* the input into the output. This transformation (which is in the "hardware" but can in principle be affected by further "software"), represents the *meaning* of this "bit" of information.

I can justify the use of this word further by reminding you that *meaning* and *intention* are inseparable aspects of one over-all activity. Intention leads to a corresponding *action*. When the element E_{ij} is "on," this corresponds to an "intention" to respond to V_α with the action flowing out in W_β. If it is "off," the intention is not active (though it is still a predisposition).

But as I indicated in my previous letter, meaning is more than intention to act. It is also *intention to display*, and to modify itself in accordance with the display. So, suppose that W_α is fed into a "display system," which in turn feeds back to V_α (self reference).

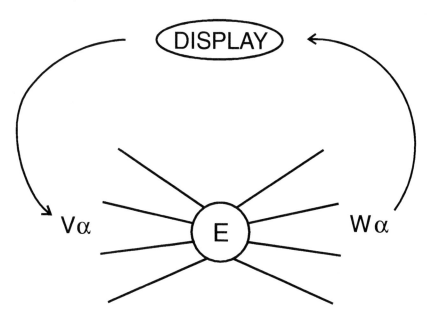

Then, if the feedback contains a structure tending to establish equivalence of V_α and W_α, we can say that when a meaning is thus *established*, $V_\alpha = W_\alpha$, and we have

$$P_{\alpha\beta} = V_\alpha V_\beta$$

(which is just the sort of projection operator that we have in quantum mechanics).

So, I say that any attempt to make something similar to intelligence in a machine would require that we put *meaning*, not just information, into the first place, in our thinking about it. Also, since every observation in quantum mechanics is basically a projection operator, we can regard this, not only as an expression of the meaning that the situation has *for us*, but also, perhaps, as a kind of *intrinsic meaning* that is present in the situation (i.e., for itself). In other words, I propose that we do not stop at regarding *information* as having an objective aspect— we should do this for *meaning* too.

[[Your computer model of meaning, intention and self reference is an interesting and instructive analogy. For me the problem remains one of finding a more direct and concrete way to approach the underlying field of intelligent perception out of which action flows and through which corrective feedback is channelled. To this end, I am especially interested in the *qualitative nature* of the transition between the limited, strictly informational regime and the full (infinitely valued) domain of meaning. How might the transformation be understood intuitively? How might its quality be brought out more concretely, most effectively embodied?

[[In his paper on "Explorations in Mathematics," mathematician Charles Muses [1977] suggests that while projection operators are normally expressed in rather abstract form, they can be given more concrete representation through use of "hypernumbers," particularly, the hypernumber ϵ, whose defining characteristics are:

$$\epsilon^2 = 1, \epsilon \doteq \pm 1$$

By using ϵ, projection operators P_1 and P_2 can be given the form $\frac{1}{2}(1 + \epsilon)$ and $\frac{1}{2}(1 - \epsilon)$, which satisfies the abstract requirements that projection operators be idempotent and that their product

be zero. Muses argues that ϵ, rather than being merely a hypo-
thetical abstraction, is an actual number provably possessing a
square root. He goes on to trace the hypernumber epsilon back
to the work of Clifford and points out that in epsilon arithmetic
the product of two non-zero numbers can be zero, thereby asso-
ciating it with the notion of nilpotence you discussed in your book,
an idea that played a central role in my interpretive essay; you
may remember my proposal that the notion of nilpotence may
be seen as a kind of arithmetic counterpart of the radically self-
referential, essentially "archetypal" form of symbolizing found
in the deepest forms of art. Moreover, Muses associates ϵ with
microphysical "spin," observing that geometrically, this would
entail a *reflection* through a fourth dimension of space which
would result, not merely in a mirror reversal (as a lower order
reflection would), but in effect, would achieve a *turning inside
out* (—your inversion experience when viewing the clown? *[see
below]*). I have suggested to Muses that the hyperdimensional
reflection he sees as underlying quantum spin would be best
represented by a Klein bottle transformation (a Moebius transformation is a reflection through *three* dimensional space which,
as you pointed out, would not be sufficient to account for what
philosopher Oliver Reiser called "circumversion": a turning inside
out).

[[There is one further element in this network of possible
connections that I believe is significant and not fully articulated
by Muses; it concerns dimension theory. Muses has implied that
the dimension theoretic counterpart of the hypernumber is not
the classical, integer dimension but the *fractional dimension*. As
I understand the implications of the latter, it entails a fundamental
unraveling of modern analytical foundations *[see chapters 4, 5,
and 12]*. I believe the "fractional dimension" must be regarded
as a non- (or perhaps, meta-) topological notion that cuts so deeply
into the assumption of analytic continuity that it leaves the
classical analyst no firm ground on which to stand. My proposal
is that there is no way to formulate properly an ϵ or Klein bottle
transformation via fractional dimensionality, as long as the
traditional stance is maintained of formulator as separate from
formulated. The discontinuity involved in this way of approaching
foundations is so fundamental that the breach can only be "filled"
by the analyst himself, i.e. by an act of radical self-reference which
closes the usual gap between the analyst and the analyzed, as

in your experience with the clown *[see below]*. In figurative terms, there is a "hole" in the Klein bottle (when the K-bottle is understood non-topologically, as a fractionally four-dimensional process structure) that the user of this symbol must fill with his *[or her]* being, just as the incompleteness of Escher's "Print Gallery" seems to be an invitation that draws the viewer into it.

[[Finally, with respect to Dirac algebra, Muses contends that the hypernumber ϵ plays a fundamental role here unbeknownst to Dirac himself. . .Would that I had the fluency to enter into a full dialogue on the specific mathematical implications of these intuitive convictions. Of course, the real issue is not a specifically mathematical one.]]

Now, to come back to symbols of self reference that actually bring the *act of self reference* about in the one who contemplates these symbols (or participates in "acting them out"), I can only wish you luck. I am not enough of a poet to do this, and I am probably a bit old to try to begin in this line. However, I would like to give you a word of warning. The repetitive use of any symbol, any ritual, or any other such activity, will tend almost irresistibly to turn the whole thing into prose (which is ultimately taken *literally*). [[In general I agree. I would emphasize that no symbol *by itself* could achieve the "poetic" outcome intended (to think that it could, would be to indulge oneself in a form of magical thinking or idolatry). I'm proposing that one would not cling to or perseverate upon the symbol as an end in itself, but would allow the symbol to merge and interpenetrate with the state of awareness being symbolized. With symbolic and meditative modes of operating open to each other, one would be working from both ends simultaneously, as it were, to bring about their conscious convergence.]]

Meanwhile, I shall give you a few ideas I have had about symbols. Firstly, in my book, *Wholeness and the Implicate Order*, (Chap. 2), I discuss the rheomode. This is basically a self referential use of symbols. Thus, consider the verb "to re-levate." As I said in the book, this itself *re-levates* a vast content, which includes the *act of re-levation itself*. So, there is a kind of self-reference here. But people have objected that it is not sufficiently poetic to grip the imagination. [[Perhaps operating in the rheomode could be further developed to make it more fully and graphically self-referential, less similar to the existing, linearly unfolded language structure that has such a powerful hold on us in our everyday dealings with the world.]]

Secondly, I recall seeing a painting by Rouault, in the Edward G. Robinson collection, entitled "The Clown." In this painting, a clown was depicted by means of a complex set of color patches, within his outward form. But on the outset, to each patch, there was a corresponding, but larger and inverted patch in a complementary color. As I watched the painting, I noticed my eyes beginning to oscillate between corresponding patches. Suddenly, the whole mode of seeing and experiencing shifted. The patches were still there, but they did not obtrude themselves on my vision. Instead, I sensed a vast flow of energy in which the "clown" poured out his whole being toward the world. This went out in a circulatory stream that twined around behind me and entered my own being. I literally experienced what it was like to be that clown. I *was* that clown, and the energy flowed in through me back to "him," to go on circulating.

Here was a symbol that was able to transform my consciousness, so that I was no longer separate from the clown that was being "observed." Whatever was true of the clown held also for me. We were both aspects of one stream of energy.

This brings me to the question of *art*, of which poetry is only one branch. I think it may be illuminating here to distinguish technology from its earlier roots, which are indicated by the Greek word "techne." This means "made by art." Obviously, technology signifies the application of "the word" (logos) to techne. This "word" may be either poetic or prosaic. In modern technology, the "word" is almost entirely prosaic (and as I said, the computer is the paradigm case of this).

Now, in a truly civilized society, the basic activity would be *artistic*. The Latin root of the word "art" means "to join" or "fit" (as in artifact, artisan, article, articulate, etc.). Nowadays, art is relegated to a secondary, aesthetic side of life, while the serious business of life is handled entirely mechanically, in a prosaic way. But in earlier times, we can see from the wide use of the word "art" (article, etc.) that art was considered as basic to the whole of life.

Now, art is not only "to join" or "to fit." For this can be done mechanically, rather than artistically. It is also to perceive new wholes constantly, wholes that are ever more encompassing, and to join or fit things according to such perceptions.

If techne were our basic activity, things would be made (joined, fitted) according to fresh perceptions of the whole. For certain purposes, this could be elaborated in a relatively fixed and prosaic mechanical way. But always, this latter would be limited by fresh perception of the whole. However, at present, the mechanical dominates, and fresh perception is allowed only insofar as it serves mechanical purposes.

If our making of things were based on techne (i.e., truly so), then even everyday life could be an ever present symbol of the oneness of observer and observed (as happened with Rouault's picture). Our houses, our streets, our structures, would all tend to lead us toward such a way of experiencing reality (whereas our present, box-like houses lead us to experience separateness and isolation). Anyway, what I wanted to emphasize is that it is *art as a whole* (and not just its poetic form) that is relevant in what you are trying to do.

[[I agree with you totally about art *as a whole*, and in using the term "poetry," I intend it in this broader sense. Furthermore, I greatly appreciate what you have been saying about art, techne and technology, and thank you for the valuable insights you provide into their root meanings. . . . Also, I am much in accord with your vision of a techne-based society, though, from my developmental perspective, I would emphasize the proposition that the state of wholeness realized by humanity would occasion a qualitative advance to a new phylogenetic level (one entailing the creative evolution of space-time-consciousness itself). At this cosmically novel and more complex level of operating, the problem of fragmentation and the challenge of achieving wholeness again would be faced (my "superman supremely ignorant" theme).]]

Yours sincerely,
David Bohm

P.S. I would like to add something about my experience with the Rouault picture.

Essentially, what I saw was that the "outside" of the clown was an *inversion* of the "inside". . . Next came the perception of a continuous flow that gradually transformed "inside" into "outside" and back again. Thus, it resembles a Moebius strip. But the corresponding figures on the "Moebius strip" are not merely mirror images, but rather, inversions of each other. This is better, as mirror imaging is too small a change to account adequately for the difference of outside and inside. Inversion is better for this purpose.

It is interesting that by extending the Dirac algebra to allow for conformal transformation, one can produce a gradual inversion, essentially through the matrices, γ^μ. Perhaps this is a clue to basic structures in the universe, which have this kind of self-reference built into them. Perhaps also inversion has its analogue in physics too, in what are called the "colors" of the quarks.

August 25, 1983

Dear Professor Rosen,

I am continuing my answer to your letter of July 25, plus enclosures, including the transcript of my previous [June] letter to you.

On page [224], you ask whether the relativism of soma and significance could be viewed as a development, entailing a "somatizing" of what was ineffably subtle. In a way, this is what happens in all our organized body of knowledge, including science. We constantly reduce subtle biological and mental processes to soma. But in so doing, we are led to deeper levels of significance. E.g., the molecular structure of DNA has the implicate order of quantum theory underlying it. And if the implicate order is eventually "somatized," something yet more subtle will underlie it.

Page [225]. The Klein bottle is better than the Moebius strip. But I think the inversion that I described in connection with Rouault's picture of the clown is still better.

Page [226]. Of course, the initial insight is "cooled down" with any words, even poetic words. Mathematical "words" tend to be much more prosaic, hence much "cooler." But mathematics can *initially* express a poetic meaning too. So you are right to look for a "poetic mathematics." But with repetition, even this would eventually turn into prose.

Poetic forms like music can embody the implicate directly. In music (as explained in my book), the over-laying of successive notes in consciousness gives a direct sense of the implicate. In poetry, the over-laying of different but related images acts similarly. In art, the Rouault picture showed me that the over-laying of "inward" and "outward" forms, through viewing in immediate succession, does the same. Perhaps it is a bit like the principle of the stereoscope, but immensely more vast and subtle.

When the over-laid images are "fused," a whole new implicate "world" comes into being. [[The research of psychiatrist Albert Rothenberg (1976) suggests that artistic processes generally entail "homospatial perception," the superimposing of two (or more) different images in the same space.]]

Page [227]. You are right to say that the *form* and *content* must agree, for an effective poetic statement.

About reality being active meaning, therefore inherently incomplete, we do have, as you say, an *actualizing* infinity.

We need to have a poetic language that directly refers to "the dance of the mind." The rheomode (in my book) does refer directly to this "dance" but not in a very poetic way. For the mind to see this, it must be in a higher state of energy than that of the ordinary mode of consciousness. A real *somatic* change has to take place in the brain and nervous system, for this to happen (chemically, physically, electrically, etc.). But that which has immense *significance* can arouse the brain to such a state of energy. The ordinary affairs of life generally lack this significance, and quite often have a negative, inhibitory significance for the somatic activity of the brain/nervous system. The whole ambience of work and home is crucial here.

Page [229]. Perhaps a poetic symbolism can mend the schism in thought, but remember the problem of repetition yielding prose. [[I quite agree that this can be a difficult problem. On the other hand, repetition sometimes can be used to deepen or expand awareness, rather than contract it, as in the case of the mantram. Perhaps the effect of repeating an outward symbolic form depends on the extent to which it is in resonance with inner dimensions of being.]]

Page [230]. I feel that the creative perceptions of intelligence are more primary than any symbolism.

Page [233]. I feel (as in the above remark about page [230]) that totality is beyond symbolizing (except as an *actualizing* infinity that is, in some sense, still relative and subject to further development). However, a genuine contact with totality should allow a person to disclose a kind of "contour." But probably, only another person who was at least beginning to go into a similar state of consciousness could really apprehend the significance of such a disclosure.

Mental activity, rising to a movement of self-actualizing infinity is, in my view, a kind of "door" through which consciousness can pass, to "enter" the state of contact. The ordinary state of mind, as I have said, cannot even approach such a "door."

Page *[234]*. Proprioception is, as you say, an incipient contact with totality (what I called a "door" above).

My own view is that the symbolic structure of thought has to "dig" to leave room in the brain for the fresh operation of intelligence and love. The "empty" brain can then create, from moment to moment, a fresh symbolic structure, as needed. This will be different from the old one in crucial ways, though similar in others. Such a notion seems closer to what happens in biological evolution than is the idea of integrating *existing* symbolic knowing and being. The knowledge that *exists* is literally *infected* with the confusion that pervades the ordinary low energy state of consciousness. [[Would *poetic* knowing or symbolizing be equally infected? Would it not, to some significant extent, directly participate in the operation of intelligence? I am suggesting that Art can be a *bridge* between prosaic, low energy symbolizing and intelligence.]]

Rather than say we may ultimately have a *synthesis* (pulling together) of knowing and being, I suggest that there is an unknown ground out of which being (which is also still a kind of knowing) and knowing (which is also still a kind of being) both unfold, inseparably related.

P. *[237]*. What you say about the incompleteness of the Klein bottle is interesting. The Rouault picture of the clown was also incomplete, in the sense that the *relationship* of inside and outside arose in the successive focussing of the eye on these two parts of the picture. As a static picture, there was no way to depict this relationship. But in any viewing, the eye (and the attention) *always* move. This was hinted at by the complementary forms and colors that were found inside and outside the clown. But the eye-brain had to follow these "hints" [[just as I suggested that proprioceptive consciousness evidently must follow the Klein bottle's "hint," fill the otherwise unfillable "hole" in this self-referential symbolic form; without proprioceptive awareness, the Klein bottle is merely an uncompletable topological structure]].

Page *[240]*. We can relativize the relativistic notion of an entirely open unreachable infinity, as you say. But this very act is implicitly an absolute. Suppose that we relativize this too. Wouldn't this bring back an open sequence of relativizations of relativity, stretching once again to an unreachable infinity? [[Yes,

I can see that, and my proposal would be that this too would be "absolutized" or closed, only to be followed by a re-opening at a still higher level, and so on. Therefore, ultimately, neither the open nor the closed, the relative nor the absolute, would prevail; rather, the perpetual *interplay between them* that I have called "relative absolutism," would forever sustain the mounting evolutional spiral.]]

Probably, we agree (as I said in connection with page *[234]*) that knowing and being interpenetrate. But knowing and being are both concepts, as is the distinction between them. So their scope is limited. In the ground, all distinctions dissolve and merge, while they yet remain distinct. That is, the ground is characterized by the oneness of opposites (e.g., the identity of identity and difference) and also by their duality (the difference of identity and difference). The ordinary state of consciousness cannot deal with this, except as a paradoxical verbal form. Perhaps a poetic symbol can bring us to the "door," and hint at it, as you say. I think that the notion of synthesis (pulling together) is too limited for this context. Rather, there is a primal whole or totality, which enfolds the opposites, and enfolds the distinction between them. The opposites are *already* together in the ground, while their distinction is also already enfolded in the ground. So they do not have to be *put together*. Rather, we have to be aware of how they actually *came apart* by unfoldment into an explicate order. In doing so, they merely manifest certain aspects of what they already are in the ground.

Is there a state of consciousness, which has died to the absolute separateness of opposites, and which starts *afresh* from the ground, in which they are one and yet distinct?

Of course, for the human sphere, I agree with your demand that there be relative incompleteness, an actualizing infinity. But this flows out of the kind of ground that I indicated above.

I think that this more or less answers the points raised in your comments on my earlier letter to you. (For those points not discussed here, you may assume essential agreement, or else, it will be evident that they are covered by comments on previous points.)

I have found this correspondence very helpful, and appreciate very much the trouble you took in making a transcript of my earlier letter. I hope that we can go on with this, and work things out further.

Yours sincerely,
David Bohm

[[I believe that in many important respects, our two positions are quite harmonious and I am encouraged by this. But it seems there are some differences of emphasis and interpretation involving the use of terms like "totality," "synthesis," and "being and knowing." The process of attempting to crystalize these differences has been difficult for me, and I recognize that the following reflections are far from the "final word."

[[I agree that there seems to be an aspect of nature that is beyond any particular form of symbolizing. I would propose further, that it is beyond any particular state of consciousness as well, that it is unapproachable in particular terms. If I were to call this aspect an "unknown ground," I would want to add that it is not only unknown to us human symbolizers but also, not even known to itself, for I view this overall feature as nature's indwelling spontaneity, the open-ended, evolutional, process aspect of nature. So the "totality" beyond any particular form of symbolizing would not be a pre-existing, already fully formed actuality, but nature's inexhaustible potential for advancing into novelty (as a Whiteheadian might say). Or we might say there is no "totality," if "totality" is intended in some sense as a final closure.

[[Of course, nature is not *sheer* openness, process or novelty. As I have proposed before, there are periods of condensation, closure or enfoldment in which ever more encompassing sub-wholes (not parts) are formed. (I think we might agree on the notion of forming ever more encompassing sub-wholes; I would only emphasize the idea that there is no already actualized, absolute totality existing *beyond* the process, but that the process *is* the "absolute totality.") Therefore, while we human beings could never concretely grasp or enfold the "absolute whole" (there is always a novel *un*foldment transcending the scope of *any* particular level of being), we can grasp the "relatively absolute" sub-whole—that which is absolute, in fact, is an infinity, *relative to* our own sphere of operations. This is precisely the challenge of actualizing infinity I believe we confront.

[[I agree that we do not "synthesize" this human totality in the sense of pulling together that which began in a state of simple separation. I concur that what we begin with is the *whole*, but suggest it is a relatively *undifferentiated* whole, a primordial whole whose enfolded distinctions are only incipient (are preconscious, with "levation" or crystalization in conscious awareness being required for their actualization). Indeed, if primally enfolded distinctions pre-existed as ready made actualities, would the operations of unfoldment and re-enfoldment (—was "reinjection" the term that was used in your dialogue with Renee Weber? [1982, p. 39]—) have any creative significance? More generally, could a simply pre-differentiated nature allow genuine creative advance?

[[In your book you spoke of creating distinctions that serve useful ends, and in my interpretive essay, I proposed that this is part of a process of differentiation in which eventually, we lose sight of the wholeness out of which the distinctions arise with conscious awareness (the result is fragmentation, as you pointed out so compellingly). The challenge then, as I see it, is to bring preconscious, primordial human wholeness (what I have referred to as "being") into intimate, interpenetrative harmony with differentiated consciousness ("knowing")—yes, they interpenetrate from the first in an incipient fashion; the goal is to bring this interpenetration to fruition, apprehend this non-duality in a conscious (dual!) manner. This is what I mean by "synthesis," and it is not so different from the goal of psychotherapy as typically stated: "to make the unconscious conscious." (In several of my papers I have spoken of "epistemotherapy" [see chapter 9], in recognition of the pervasive societal and planetary fragmentation that appears to warrant going beyond the purview of conventional therapy, limited as it is to *individual* healing.) I suggest that this "therapeutic synthesis" might be achieved through the poetic, deeply proprioceptive symbolizing/meditating we have been discussing, not unlike C. G. Jung's proposal that the individual can achieve a therapeutic centering through archetypal awareness attained by active imagination. (Note that while my proposed extension of active imagination to a communal scale would not encompass the "unknown ground" or "absolute whole," it would lead to a concrete grasping of our human sub-whole through poetic symbolizing/meditating of a ("sub-") universal nature; interestingly, Jung himself, toward the end of his life, saw the need for tapping the most general forms of archetypal expression [as noted in chapters 9 and 12].)

[[Let me try to summarize my position in relation to your response to my comments on page [230]. You suggest that "totality is beyond symbolizing (except as an *actualizing* infinity that is, in some sense, still relative and subject to further development)." My own feeling is that we can speak of totality *only* as an actualizing infinity, as a *relative* absolute. Here I am proposing that there is no absolute totality or pure intelligence that cleanly transcends the directed, symbol-generating processes of nature. So, rather than speak of grasping totality or wholeness by emptying myself of, or entirely detaching myself from symbolic process, I would speak of directly and consciously (non-simply or dually) realizing the non-duality of wholeness and symbolic structure, with said realization eventuating in the creation of a novel sub-whole. It is as such a poetic merger (a "synthesis" or "reflexive integration") of preconscious with conscious awareness that I would understand the creative perception of intelligence.]]

[Handwritten post script added by Rosen:] [[In the context of responding to the idea of a pre-existing absolute totality, I may have put a slightly greater emphasis on the open aspect of nature than on the closed. I hope it is clear that the notion of "relative absolutism" does *not* give primacy to openness but suggests the ultimate non-duality of openness and closure. So if nature is forever opening, just as much, it is forever closing, and if no closure or totality is absolute, neither is any state of openness, partitiveness or incompleteness—there is always an *actualizing* of infinity, however great the fragmentation may seem to be.]]

OCTOBER-NOVEMBER

[ca. October 15, 1983]

Dear Professor Rosen,

Thank you very much for typing up my last two letters, and including your comments.

As I see it, the key issue is that of the nature of infinity. I include a new paper on soma-significance, which should help throw some light on this question.

For the present, I would only say that it would be best not to put any restrictions on infinity at all—not even that it has to be "unformed" and "potential" or "self-actualizing and self-closing." The word "infinity" means "no limit, qualitative or quantitative."

As I see it, the key question is: What is the relationship of the infinite and the finite—i.e., that which is, in one way or another, limited? We can *propose* various ideas about this, "play" with them, criticize them, and eventually, if they stood up, we can propose to live by them, unless and until we meet contradictions—in which case, our ideas would have to change. We also have to be alert as to how fruitful it actually is to live by the ideas in question.

Apropos of the finite, it is interesting that "reality" is based on the Latin "res" meaning "thing." But this is derived from "reri,"—to think. At this rate, "the thing" is "what we can think about." Whatever we can think about is *limited.* For thinking requires determination, and to determine is to limit—i.e., to "terminate." To say that a thing is X is also to say that it is not "not-X." So all determination involves negation—hence limitation. If it has to be X, it cannot be "not-X." I would say that infinity, by its very essence, cannot be limited to any X or not-X. It may *limit itself* to X, but this cannot be guaranteed or universal. That is to say, the total activity of *all limitation* is contained in the infinite. So, if you say that infinity is limited to X, you are implying that it is limiting itself to X. But to this, I would say: How do you know? How do you know that in other contexts as yet unknown, it also limits itself in this way? What we have been discussing is the extension of the *finite* into new domains. To do this is, we hope and expect, likely to be fruitful. We may consider, for example, the "eternally opening and closing infinite." But this is still limited to "eternally closing and unclosing." So, ultimately, it is finite, i.e., limited in *some* way.

Finally, the purpose of language is to *signify,* i.e., to "point to" a reality, beyond itself but including itself. It is what I have called elsewhere part of the "dance of the mind." We are constantly experimenting with new "art forms" in carrying out this "dance." That is, at any rate, how I see it. To make our metaphysics coherent and to change it constantly so as to keep it coherent or extend its coherence is essential for order in this "dance of the mind."

Yours sincerely,
David Bohm

November 10, 1983

Dear Professor Bohm:

Let me begin by saying that I very much appreciated your suggestion to David Griffin that I be invited to participate in the March conference on "Physics and the Ultimate Significance of Time," in Claremont, California *[see chapter 14].* From David's description of the meeting, it was not entirely clear that you will be delivering your paper in person. Needless to say, I certainly hope you will! *[Prof. Bohm did attend in person.]*

Thank you for sending me your new and interesting paper on "Soma-Significance," along with your covering letter. Regarding the latter, firstly, I agree that the key issue for us concerns the nature of the infinite, the finite, and their interrelations. The essential notion I have been proposing is a *non-duality* of the infinite and finite. This results in a position I have called "relative absolutism," one that I suspect is not so different from your own position, at least as I interpret certain aspects of it that I take to be significant. Both simple absolutism (in a positive way) and simple relativism (in a negative way) would regard the infinite from *the perspective of the finite*, i.e., from within the domain of mere thinking (I agree with your assertion that thinking is fundamentally a function of the finite). While the simple absolutist mistakenly attempts to put fixed constraints on the infinite, capture it within the finite categories of thought, the relativist realizes that this *cannot* be done but can find no mode of functioning other than thought, and therefore, being strictly limited to the finite, must conclude that the infinite is forever beyond reach, utterly unapproachable. By contrast, in "relative absolutism," one may achieve a kind of "closure on the infinite," not by attempting to reduce the infinite to the finite expressions of thought, but by operating "from within the infinite," as it were, through a meditatively *intuitive* mode of awareness. Actually, as I have suggested before, such a mode of realization would not entail a transition "from" the finite "to" the infinite, but an apprehension of their non-duality (while *at the same time* grasping the fact that they are different, not simply reducible to one another), for to view the finite and infinite merely in terms of their separation is essentially, to view them in terms of the *finite*. On the other hand, when operating "from within the infinite," one's perspective is radically unitive, involving the higher-order unity of the unitive (infinite) and separative (finite). In Jungian terms, the unity to be achieved would be between the thinking-sensing function (which I believe is of the finite, the separative) and the feeling-intuiting function (which I propose we can associate with the infinite or unitive). (Currently, I am exploring the viability of a Jungian/Bohmian/Whiteheadian treatment of the problem of symbolizing in science and art.) What I have been suggesting in our correspondence is that achieving this higher-order "infinity of the infinite and finite" is our present evolutionary challenge, and that should we be equal to it, we will not have captured the infinite once and for all (as an absolutist might believe), but will confront as evolved beings, a new, more complex order of *finitude*, a new evolutionary challenge at a higher (more comprehensive and multi-faceted) level.

In any event, on the general and basic question of whether anything at all can be said to qualify authentically the concept of the infinite, a negative answer, in effect, should render us mute on the subject from that point onward. Simply to assert that the infinite is unqualifiable would seem to make *any* discussion of the matter an exercise in futility. True, if our discussions were limited merely to thinking, futility would be inevitable; we would be trapped in the endlessly self-negating regress of our own verbiage. But I believe the beauty of our exchange is that it is strongly *intuitive*, and so I feel we *can* meaningfully approach the idea of the infinite, break out of the "Hall of Mirrors" depicted by Orson Welles in his "Lady From Shanghai."

Turning to the "Soma-Significance" paper, I found it extremely stimulating and thought- (intuition-) provoking. (Happily, this is not unusual for the material you kindly have been sending me.) I will take some additional time to evolve my responses, then transmit them to you.

Looking forward to the meeting-in-the-flesh we at last may have in March '84, I am yours sincerely,

Steve Rosen

14

Time and Higher-Order Wholeness:
A Response to David Bohm (1984)

INTRODUCTION

To begin, let me thank David Bohm on two counts: first, for the correspondence we have been engaging in for a year and a half now—I have been enriched by it; second, and more important, for the courageous stand he has taken and the creative leadership he has shown in contributing, not merely to advancing the knowledge of his scientific discipline, but to the broader and more crucial quest for social healing so badly needed at this particular juncture in the history of our species. In my correspondence with David Bohm I have felt a deep harmony of interest and purpose, and the many views we have shared have been a source of gratification to me. We also have discussed some differences in perspective with an honesty and openness that I believe have been fruitful. My commentary on his paper for our conference [Bohm 1986] is made in the context of our ongoing dialogue.

In his book, *Wholeness and the Implicate Order*, Bohm observes that "to be healthy is to be whole," pointing out that "the word 'health' in English is based on the Anglo-Saxon word 'hale' meaning 'whole'" (1980, p. 3). There is little doubt that our quest for healing is a search for ways of overcoming fragmentation and achieving wholeness. But this formidable challenge is complicated by the fact that the fragmentation we face is a "fragmentation squared," so to speak. That is to say, the schism that needs to be mended is no simple division of a whole into parts, but rather, is a *higher-order* division in which the "parts" themselves are dividedness and wholeness. While the origin of this currently pathological state of bifurcation long predates the Renaissance, the condition is perhaps most clearly illustrated in the philosophical stance of René Descartes, a position that has had a profound and pervasive influence on subsequent Western thought, as Alfred North Whitehead (1978), among others, has amply acknowledged. The "Cartesian split" is in essence a division into that which is divisible—extended in space thus

differentiable, subject to indefinite analysis, precise delineation—
and the indivisible—unextended, undifferentiable, thus seamlessly and unanalyzably whole. The domain of the divisible is that
of extensive continuity undergirding all of Western science,
mathematics, and technology; it is a domain of sheer juxtaposition, purely external relations, as Milic Capek (1961) has demonstrated; in the language of David Bohm, it is the realm of explicate
order. On the other hand, the indivisible world is *dis*continuous,
strictly "atomic" or discrete, an impartible unity, a realm of pure
*in*ternality of relation, an "enfolded" or "implicate" order, in
Bohm's terminology. And by virtue of the *higher-order* nature of
the Cartesian split, the differentiated and undifferentiable regimes
are themselves differentiated, related in a wholly external manner.

As I see it, the problem of *time* is inextricably linked with
that of the higher-order differentiation. I propose that the mystery
of process or becoming, from which the familiar idea of time is
abstracted, is the mystery of how the undifferentiable becomes
differentiated. Time, I suggest, entails a differentiat*ing* of that
which is undifferentiable. Or, in more common parlance, we may
speak of an actualizing, a bringing into extensive continuity, of
the purely potential or unextended. The emphasis I have placed
on the "ing" suffix in the term *differentiati*ng is to indicate that
the undifferentiable and differentiated aspects of time must be
understood as related in the most intimate, interpenetrative way.
The "paradox" of time is that two distinct features are required,
yet these features need to be recognized as one and the same!
When they are not, time decomposes into antithetical forms of
timelessness (as I will demonstrate). In general terms, I am
suggesting that time in its essence is radically non-Cartesian in
character and that, therefore, achieving an authentic comprehension of time-as-becoming should entail a perception of higher-
order wholeness that could well be therapeutic in its effect. With
this in mind, I would like to consider Bohm's approach to the
question of time, as well as that of Whitehead. I will conclude my
presentation by crystalizing the notion of higher-order wholeness
in a graphic fashion.

"VERTICAL IMPLICATE ORDER" AND "UNKNOWABLE TOTALITY":
TWO FORMS OF TIMELESSNESS

I find myself more or less in agreement with David Griffin's
[1986b] assessment of what he calls the "Thomistic and Vedantist

moods" of David Bohm. I concur that the former is in evidence in the "vertical implicate order," while the latter shows up in what Bohm refers to as the "infinite totality," "the holomovement" *[1986, p. 200]*, or the "unknown" *[1986, p. 207]*. And it appears that both moods end in a denial of the reality of time, though on opposing grounds. Yet both moods *are* needed, since an affirmation of time in the sense of higher-order wholeness I have adumbrated entails their *integration*, as will be seen. It is here that Bohm assumes a position of leadership in his largely "Thomistic" field. Moreover, in some of Bohm's writings and in our correspondence, I believe there are indications that integration has already begun; I will return to these in the concluding part of my commentary.

If Griffin's interpretation of Thomas as espousing a metaphysics of the finite, of wholly differentiable form, is correct, the complete *relativization* of the implicate, enfolded, undifferentiable order found in "vertical implicate order" does seem to be Thomistic. That which is implicate ("infinite") in one context is, at a deeper level, in a more comprehensive context, *explicate* (finite). Concomitantly, that which is indeterminable at one level is fully determinable at the next, a notion that seems to render superficial our ideas of freedom and creativity and raise legitimate concerns among Whiteheadians in particular. Similarly with respect to time per se: "vertical implicate order" appears to divest time of its genuinely undifferentiable aspect, producing the species of timelessness that Bergson, Capek, and other process-oriented philosophers have decried—a spatialization of time.

The issue of time has been brought into focus in an especially significant arena for contemporary thought, that of modern physics. Bohm identifies the problem by pointing out that, in the standard approach to quantum mechanics, time has meaning only in relation to an observer who is outside the system. Yet the observer is intimately linked to the system. This raises the spectre of the subjective—of the inherently undifferentiable Cartesian *psyche* or mind—in the very midst of "objective," differentiable *physis*. And it is this that Bohm, in his "Thomistic mood," wishes to avoid. In order to account for the subjectivistic time associated with the observer in a strictly *objective* manner, this concretely undifferentiable implicateness is embedded in a deeper, *abstract* order by invariant algebraic transformation, and it is thereby relativized, rendered explicate, made timeless in the sense of the spatialization of time. So because time cannot be accommodated by concrete physical continuity in the quantum-mechanical system, Bohm introduces a mathematical continuum or "space" whose "dimensionless points" are invariant algebraic operations. In general, it appears that "vertical implicate order" is an indefinitely extended hierarchy in which the appearance of undifferentiability in one context is countered by abstractly invoking a more comprehensive context. Rather than facing the problem of undifferentiability head on—in classical Cartesian language, facing the problem of mind—this problem is put off interminably by an endless sequence of abstractions upon the differentiable or physical. Therefore, "vertical implicate order" does not come to grips with the genuinely, concretely undifferentiable feature of time, the truly atomic aspect of process.

Of course, Bohm is well aware of this limitation. He *admits* that such mathematical abstractions are only analogies that do not in themselves approach the absolute implicate order, the "infinite totality." And this brings us to the so-called Vedantist mood of David Bohm. According to Griffin, Bohm sometimes speaks "as if the ultimate reality, the ultimate implicate order, were totally formless." In such a case, "all form and measure would be *maya*, illusion" *[Griffin 1986b, p. 137]*. At times in my personal exchanges with Bohm, I too have gotten the impression of an ultimate denial of form in favor of that which is formless. For example, he has distinguished symbolic knowing from what he believes to be beyond *any* form of thought. Bohm *has* acknowledged that certain forms of symbolizing may usefully call attention to their own limitations and therefore serve as stepping stones, paving the way for transcendence. But in the end, through the acts of inward awareness and deeply reflective attention, which are *distinct* from mere forms of thought, form is entirely left behind; it dissolves in an "intelligent perception of the infinite totality." As I see it, the nonduality thus achieved preserves the higher-order *dualism* of the finite and infinite, the differentiated and undifferentiable, for by granting formless totality such priority over form, form does not merely vanish but remains to express itself negatively in the now unsolvable enigma of why there is form at all.

If I am correctly interpreting the paper Bohm has presented for this conference, the same strain of "Vedantist dualism" (along with the enigma it implies) can be detected in the concluding portion *[Bohm 1986, pp. 196ff.]* and is particularly evident at the very end of the essay in connection with our central question of time. Here Bohm states: "the clear perception that we are the unknown, which is beyond time, allows the mind to give time its proper value, which is limited and not supreme. . .What we must inquire into deeply at this conference is, then, the inverse question of what the significance of time and thought is for eternity. This is a very difficult question" *[pp. 207–8]*. Indeed, it is the "Vedantist" enigma I have referred to. As I finished my reading of the paper with this statement, I was struck by the dualism inherent to its structure and found myself writing in the margin, "But isn't the whole point about wholeness a point about the *non*duality of time and eternity?" Note that whereas the view of time I am proposing requires that time possess nondually, both a differentiated and undifferentiable aspect, from the "Vedantist" standpoint, "time" is identified with the strictly differentiated and "eternity" with the wholly undifferentiable, thereby giving the second kind of timelessness, the one antithetical to the "Thomistic spatialization of time."

THE "ACTUAL OCCASION"

The third corner of the "triangle" for this conference is the philosophy of Alfred North Whitehead. Let me give my current understanding. The Cartesian assumption of *extensive continuity* appears at the deepest level of Whiteheadian metaphysics (see, for example, Whitehead 1978, p. 66). As I observed earlier, the relation that obtains between events in an extensive continuum generally is one of mutual externality. The way Milic Capek has put it, in a simply continuous space "no matter how minute [an] interval may be, it must always be an *interval* separating two points, each of which is *external* to the other" (1961, p. 19). To be sure, in Whitehead's philosophy of process, the fundamental unit is not spatially static but dynamic; of course I am referring to the "actual occasion." Each "actual occasion" is seen as an event that entails the "prehension" of the objective world by the self-determining subjectivity of that moment. In the enfoldment, each "actual occasion" does overcome the separative, simple continuity of dimensionless points abutting in space; with every occasion, extensive space is atomized, the continuous yielding to the discrete. However, the given, momentary occasion itself is *separated* from subsequent occasions in the higher-order, light cone-constrained process continuum that is presupposed. So while every occasion renders the differentiated undifferentiable, the occasions themselves are *differentiated*, external to one another. One way this is expressed in the development of Whiteheadian thought is by the idea that, when the "concrescence" of a given occasion is complete, the subjectivity of this moment has utterly "perished," that is, the occasion has been totally objectified or "superjected." In the succeeding occasion, what had just comprised an inward, living subjectivity is now simply external and lifeless, immortalized merely in objectivity, an aspect of the external world to be shaped afresh by a *newborn* subjectivity. The Whiteheadian "rule of thumb" is stated by David Griffin in his contribution to this conference: "subject *then* object" *[1986b, p. 141; emphasis added in my original paper]*. But to grasp the true working of process, is it really enough for us to view subjectivity merely as prehending other-as-object, that is, *after* the fact of this other's own living subjectivity? Would not genuine transition demand the *trans*subjective, core-to-core immediacy of presence of that which is other?

In terms of our focal issue, I venture to say that the Whiteheadian assumption of extensive continuity blocks the attainment of the radically non-Cartesian, higher-order wholeness that I believe is necessary for an authentic comprehension of time as becoming. To reiterate what I see as the fundamental "paradox" confronting us, time is a differentiat*ing* of the undifferentiable. Thus, two distinct aspects are required, the differentiated and the undifferentiable, yet they must be related in a profoundly intimate fashion, viewed not merely as "sides of the same coin" or "end poles of a single continuum" but as interpenetratively one and the same. My understanding is that Whiteheadian metaphysics stops short of this "nondual (monistic) dualism" *[chapters 3, 10, and 11]*. The consequence is that, in a subtle way, again we seem to be given an admixture of "Thomism" and "Vedantism." The continuous and the discrete, the well-formed and formless, the differentiated and the undifferentiable are brought into simultaneous view, positioned in close proximity, placed back to back. But they are not convincingly integrated, not fully reconciled, so enigma persists. The mere alignment of opposing forms of timelessness—a completely determined "Thomistic past," an utterly indeterminable "Vedantist future"—does not give us time.

Time, I maintain, is the *nondual duality* of the determined and indeterminable.

THE PHENOMENOLOGY OF TEMPORAL STRUCTURE

At this point I would like to turn away from a purely conceptual approach so as to bring the notion of 'monistic dualism' into concrete, graphic focus. My intention is to attain a closer agreement between form and content. Although abstractionism may be a natural form of discourse when the content is Cartesian, I believe the subject matter of the non-Cartesian approach I am advocating additionally demands experiential immediacy for full justice to be done to it. And this should be in keeping with the frequent recommendations of David Bohm, who has been a leader in the struggle for recognizing the prime importance of direct intuitive understanding in theoretical science. I will begin with a phenomenological exercise and segue into what might be called "poetic mathematics," an expressive form to which Bohm also has contributed . . .

[*Here, as in earlier chapters, the Necker cube is employed. Note that in the original essays on which chapters 1, 3, 8, 10 and 11 were based, the integrative quality of the cube was brought out by comparing it with a* divided rectangle. *In this present book, said comparison is made explicit only in figure 1.4. In the case of the present essay, the cube is again compared with a divided rectangle, the two separate parts of which we now imagine to be labeled "D" and "U," to signify the differentiated and the undifferentiable, respectively. When visualized as the overlapping perspectives of the cube, D and U are pictured in more intimate relationship with one another.*]

The proposition I offer is that the divided rectangle be taken as illustrating simple Cartesian structure while the Necker cube—in our *customary* way of viewing it—be interpreted as exemplifying the subtler, temporal dualism evidenced in Whitehead. Philosopher David Haight (1983) has argued that Whitehead's dipolar monism is a clear improvement over Descartes's dualism, yet still too dualistic to capture "the absolute and actual [as opposed to merely potential] simultaneous presence" of actual entities to each other (Haight's approach is not dissimilar from my own, though there are also areas of difference beyond the scope of the present discussion). *U* and *D*, the two alternating perspectives of the cube, may be seen to represent the "mental" and "physical" (or "atomic" and "continuous") "poles" of White-headian metaphysics, respectively. True there is no *spatial* separation of poles, as there is in Cartesianism, but poles *are* rendered disjoint by assuming a relation of simple succession. Again, the Whiteheadian rule is "subject *then* object," and in this disjunction interpolated between poles, a "quantum leap" must be taken from one to the other, leaving us incognizant of what happens in between. In this sense, poles are *completely* polarized, related in a strictly external manner. Interesting and relevant in this regard is Bohm's definition of a mechanistic relation: "entities. . .*are outside of each other*. . .and interact through forces that do not bring about any changes in their essential natures" (Bohm 1980, p. 173). Applying this definition to White-headian dipolarity, we may observe that no changes in the essential natures of the continuous and the atomic are brought about by their purely successive interactions. The continuous remains simply continuous or differentiable; the atomic remains simply, undifferentiably that.

Now let us go a step further in our phenomenological exercise with the Necker cube. . .*[At this point, the* perspectival integration *of the Necker cube is enacted (chapters 5, 6, 10, and 11) as the means of symbolically surmounting Whiteheadian dualism.]* Thus, the "one-sided" entity produced in the visual exercise with the cube may be said to exemplify the relation of *monistic dualism* required for reconciling the differentiated and the undifferentiable. The result is an insight into the hidden structure of time, the "crack" between simultaneity and succession that is neither alone but both—in fact, nonmechanistically *more* than both, the more being none other than movement, process, becoming itself.

In our final exercise, the proposed solution to the problem of time will be delivered in still more concrete form, the structure being tangibly "materialized," so to speak. We turn to the field of mathematics known as topology, for here a palpable model of one-sidedness can be fashioned. But let me repeat that my intention is not to develop a rigorous mathematical formalism, but to go beyond "Thomistic" rigor to a "*poetic* mathematics." So we will start with topological analogy, then advance—not to axiomatically airtight homology but in the other direction, to "airy" metaphor, though it be mathematical nonetheless. . .

[The "topological analogy" that is offered is, of course, the surface of Moebius.]

But the Moebius strip *is* only an analogy. It *symbolizes* the union of the differentiated and undifferentiable but is itself simply extended in space, subject to precise delineation, wholly differentiable. Now Bohm has written eloquently on the importance of poetry and art in our search for healing (= wholing), the importance of *metaphor*. In our correspondence, he has pointed out that while mere analogy (simile, prose) does not give us direct access to the implicate, metaphor *can*:

> A metaphor is an *implied comparison*. It is to be distinguished from a *simile*, which is an explicit comparison. In a simile, one says A is similar to B in certain explicit respects, different in others. The elaboration of a simile ultimately gives rise to *prose*, which is an *indefinitely extended simile*, i.e. to talk prose is to say (or imply) that the world is *like* what I say (though also *unlike*). In a metaphor, it is essential that the similarity is left implicated or unstated. I suggest that *the meaning of a metaphor is therefore directly apprehended in an implicate order of the mind*. So in poetry, it is possible to engage the implicate "dance of the mind." *[see chapter 13, p. 225.]*

Bohm's distinction between simile (analogy) and metaphor can be seen as providing a basis for a "poetic mathematics." The key is that, in such a form of expression, "similarity [would be] left implicated." In other words, a "poetic mathematics" would not be predicated on the effort rigorously to delineate, explicate, render invariant; it would accept the radically *non*linear, the implicate or noninvariant, in its very midst. From this standpoint, "vertical implicate order," understood as a hierarchy of invariant algebraic transformations, would have to be regarded as prosaic. But elsewhere Bohm emphasizes another aspect of algebraic representation. He introduces the idea of "nilpotence," that is, nonzero terms that when multiplied become zero! Bohm observes that "properly nilpotent terms describe movements which ultimately lead to features that vanish [rather than ending in well-delineated products]" (Bohm 1980, p. 169). When our aim is invariance, nilpotent terms must be excluded, says Bohm. Conversely, the inclusion of nilpotence should serve the goal of "non-invariance" or "impermanence," that is, of a metaphorical movement that can directly engage the "dance of mind." In my essay on Bohm's book *[chapter 12]*, I noted that aspects of mathematician Charles Muses's "hypernumber algebra" (1977) seem to serve the same purpose and are closely related to the notion of nilpotence. Moreover, I believe it can be shown that the topological counterpart of the hypernumber epsilon[1] is the *Klein bottle*, the latter being none other than the higher-order version of the Moebius surface, the "metaphorization" of the Moebius analogy. I propose that the Klein bottle—an "inside-out" structure that *literally* goes through itself—delivers the Necker-cube expression of time-as-authentic-becoming in directly implicate, poetic fashion.

Of course, it is possible to treat such poetic mathematical forms in a prosaic way, to reimpose invariance by the use of abstraction. Indeed, the deeply engrained "Thomistic" habit is to do just that. As Capek has pointed out, the spatialization of time is an almost irresistible tendency for the post-Renaissance mind. We simply cannot bring ourselves to *accept* impermanence, noninvariance, the flux of process. Instead, we attempt to complete the incomplete by an act of mere abstraction that takes us but another step in an infinite regress. I propose that we can complete incompleteness only by genuinely *accepting* it. Here the "hole" in the Klein bottle (construction of the bottle cannot be completed in three-dimensional space) would not be filled by abstract spatialization, as is the topological convention. Consciousness itself—what Bohm calls "intelligent perception"—would fill this hole. In this way, "Thomism" would be integrated with "Vedantism," the differentiated reconciled with the undifferentiable, the net result being *process*.

So in the final analysis, I believe there is much common ground between David Bohm and myself. I would only emphasize my conviction that "Thomistic" and "Vedantist" traditions cannot properly be unified if poetic form is considered a mere stepping stone to formless intelligent perception, to be left behind when the "infinite totality" is realized. In the "poetic mathematics" of which I speak, form and formless intelligence are dually one and the same. It is this monistically dual relation that underlies the paradox of time as process, change, creative advance. Therefore, we may view it as the key to our freedom.

Epilogue

The Limitations of Language and the Need for a "Moebial" Way of Writing

The chapters of this book trace the evolution of the *Moebius principle*, a new way of approaching the foundations of science and philosophy. The strategy has been to confront crisis and fragmentation in contemporary thought by offering a concrete intuition of thoroughgoing wholeness. In concluding my work, I will give a capsule summation of the main features of the Moebius intuition, then briefly foreshadow the current trajectory of my ongoing enterprise.

In the Moebius principle, wholeness is sought in the embodiment of paradox. This provides a "higher-order wholeness," that is, a wholeness of wholeness and separateness in which the two are fully integrated yet also entirely distinct (they constitute a *nondual duality*, not a mere nonduality). The wholeness in question is utterly fluid and dynamic, an unobstructed boundless flow. It entails a transformation of identity or self that is self-enacted (not externally induced). Finally, the Moebius process is *radically* processual, in the sense that it involves *multiple levels* of process, with processes at any given level *themselves* being processed (as brought out via the "Moebius-Klein system" [or "generalized Moebius sequence," as it is called in chapter 5]).

Over the eighteen-year period during which the essays of this book were written, I have gradually come to recognize a basic limitation of my work, and this has affected the emphasis and direction of my efforts (as indicated in some of the more recently written chapters, such as 6 and 11). To put the problem succinctly, if I am seeking wholeness in the fullest meaning of the word, it is not enough for me merely to write *about* it; wholeness must be embodied in my *own way* of writing. The basic impediment here is that writing is an activity whose conventional form serves to *obscure* wholeness. Though this concealment is epitomized in the scientific and mathematical form of writing, the underlying problem is neither scientific nor mathematical per se, but concerns the nature of language in general and as such. In arriving at this realization, my focus has broadened from its previous preoccupation with the foundations of science and mathematics to the more general philosophical matter of language (and thinking).

It is not that this matter was entirely ignored in my earlier work. Indeed, the questions of language and symbolic activity in general figured prominently in my dialogue with David Bohm (1982–84), as evident in the foregoing section. Nevertheless, even at times when I did address the limitations of ordinary language, when the need for a more *poetic* mode of discourse was expressed in what I wrote, essentially, it was still a *writing about*. What I have come to understand is that, if I am to write wholeness (not just write *about* it), beyond expanding the scope of my subject matter to language in general, my own manner of writing needs to be less prosaic and more poetic, more meditative or "proprioceptive" (as Bohm would say). Whereas prose *objectifies*, leaving the subject entirely implicit thus perpetuating the total splitting of object from subject, poetry allows the subject to reverberate amidst the objectification, and a *fully* meditative or proprioceptive writing (one that surpasses *traditional* poetic forms) would integrate subject and object in a complete and direct way.

So my writing must be thoroughly proprioceptive, it seems. And this implies that I, the writer of this book (the subject who has created this object appearing before you), would need to bring *myself* into my writing far more than I have done in the past. As long as I cloak myself in authorial anonymity offering my reader naught but an object, wholeness will not be forthcoming. It is *myself* that I must offer to my reader. Does this mean *describing* myself to you, say, through an account of my personal characteristics and life history? Surely it cannot simply mean that, for in any act of mere de-scription, this subject would be representing himself through the disembodied medium of language that would only objectify him. Evidently, I somehow must bring subjectivity into the text as *prelinguistically embodied.*[1]

Let me try to approach this difficult matter a little differently. To heal the split between subject and object, to put subjectivity into this text without turning it into still another subject-denying object, something has to happen to these very words I am writing, **these words printed before you.** I must apply the Moebius principle to *them*.

In ordinary writing, these bodies of ink are objects that stand in for and give expression to my subjectivity in such a way that *living* subjectivity is hidden (even when I explicitly refer to myself as the subject who writes). Here, all we encounter are these linear configurations on the page—arbitrarily devised and conventionally agreed upon signifiers with no intrinsic meaning of their own, thus bodies which are lifeless—and the disembodied meanings they signify. It is the Whiteheadian "rule of thumb" that applies in ordinary writing: "subject *then* object" (see the previous chapter), so that, the instant the object appears, the living subject seems to have perished completely. Yet, applying the Moebius principle, I suggest that prelinguistic (preobjectified) subjectivity actually *continues to operate* within these seemingly fossilized linguistic objects (in Gendlin's terms, the "implicit intricacy" continues to function in the midst of the "forms and distinctions" created from it [1991, p. 37]). In the Moebius way of writing, this ever-present wellspring (this subjectivity that precedes the split between "subject" and "object") would be made transparent or diaphanous.[2] In forming these words, I would recognize them as products now being projected from "somewhere deeper or earlier," and, proprioceptively stretching my attention further backward into the "hole" or "blank spot" in the field of language (Rosen 1993), I would re-member the fleshy womb of these skeletal signifiers, the living process from which these "lifeless" objects are coming into presence. Writing in this way, no longer would I be engaged in *self-concealing* acts of projection. I would be writing with a primordial, inside-the-moment awareness that would "intercept" these abstractive projections of language, apprehend them "before" they are sent forth from the prelinguistic bodily matrix in which they gestate.[3] Such a proprioceptive "unwording" of the words I write would be shared with my reader. And if the reader too would "moebiusly" stretch her or his attention backward to the "inner horizon" of the word thereby approaching its wordless origin, the word might cease to be a devitalized means of communication that blocks our bodily communion; it might become a "*living word*" (to use the term of literary theorist Christa Wolf).[4]

In sum, it is my present conviction that the Moebius principle must encompass a new way of writing, one that goes beyond the objectification of wholeness to encompass the deeper wholeness of object and subject. Moreover, this radically holistic mode of expression would require and invite reciprocation on the *reader's* part; its effectiveness would depend on *us*.

I hope that these closing paragraphs have managed to convey some sense of what now concerns me. Be that as it may, I can go no further, since the aim of this epilogue is to conclude the present book, not to begin a new one!

NOTES

PREFACE

1. Perhaps, since I will be drawing on Heidegger in several of the chapters to come, I should take this occasion also to acknowledge the controversy surrounding his affiliation with Nazism in the 1930s. There has been a great deal of heated debate over this complicated matter. Here I must limit myself to stating only my own conclusions without going into detail.

   Beyond Heidegger's political naivete and weakness as a person, I can understand how Heideggerian philosophy itself might be distorted to serve the cause of the Nazis. And yet, at the same time, I can think of no major Western philosopher who saw more deeply into the roots of Nazism. Nazism can be said to stem from an oblivion to wholeness that Heidegger was seeking to surmount, especially in his later work. In Heidegger's own words: "In . . . 1934–35 I gave my first Hölderlin course. In 1936, I began my lectures on Nietzsche. Anyone who could hear, heard that the latter were opposed to Nazism" (1977e, p. 13).
2. Recently, I learned of two attempts at deconstruction that are in some respects similar to my own: Gendlin and Lemke (1983) and Schumacher (1989). Both of these works show the influence of phenomenology à la Heidegger and Merleau-Ponty, but they also manage to operate *within* the foundations of science, at the inner core of science, in the field of modern theoretical physics (rather than being limited to more peripheral fields). Thus they do more than provide a merely external philosophical critique.

CHAPTER 1 (1975)

[Following its appearance in Main Currents, *this essay was further revised. The present version takes into account both the published form of the essay and the revisions done not long after its appearance in 1975.]*
[1. The reader may find it interesting to note how gender biased language disappears in my later essays.]
2. Georgescu-Roegen's sense of this term will henceforth be used.
[3. Note that the operation of turning a single profile into its opposite, depicted in figure 1.2, is clearly equivalent to that of superimposing

273

a pair of oppositely oriented profiles. In both cases, the separation of right and left is surmounted by hyperdimensional rotation.]
[4. See chapter 5 for a more detailed and sophisticated treatment of the Moebius strip, the Klein bottle, and their interrelationships. Note also that additional insights on the Klein bottle are given in chapter 11.]

CHAPTER 2 (1975)

[1. In the present exercise, the reader is required to picture the profile as drawn into the surface, rather than positioned upon it.]

CHAPTER 3 (1980)

[At the time I wrote this essay, I was unaware of the famous work by Henri Bergson bearing the same title.]
[1. In chapter 1, the words "Being" and "Becoming" are capitalized. This reflects the more metaphysical connotation of these terms in the earlier essay.]

CHAPTER 4 (1983)

1. Henceforth, when the term *continuity* appears without being qualified, it will be understood to mean *simple* continuity of the analytic kind, not the more general form of continuity I intimated earlier, in speaking of "deeper" continuity. The latter notion will be considered again in the final section of this paper.

CHAPTER 5 (1988)

1. Capek (1961) came to a different conclusion concerning the implications of relativistic space-time. The limitations of his argument have been examined by Graves (1971, pp. 249–50).
2. The distinction that I have drawn between the "internal" or "intrinsic" character of space and its "extrinsic," topological structure should not be confused with the strictly *topological* distinction between "intrinsic" and "extrinsic geometry." Both of the latter terms refer to the global structure of a manifold—that built up from structureless identity elements. In the "intrinsic" case, global properties do not depend on the space in which the manifold is embedded, while, in the "extrinsic" case, they do (the orientability of a surface is an example of an intrinsic feature, and the sidedness of a surface is an example of an extrinsic one). In the *present* exposition, the term *intrinsic* refers to *local* relations in the manifold, that is, to the status of the identity elements themselves (as opposed to the global structures built up from them); and the term *extrinsic structure* is understood to subsume both "intrinsic" and "extrinsic geometry."

3. The idea of "fractional dimension" should not be confused with Mandelbrot's (1977) concept of "fractal dimension." In the latter, topological dimension is not affected by fractal transformations. Thus, the classical intuition of dimension is not directly challenged.

4. To sustain the neo-intuitive initiative, references to the "density values" of dimensions need to be intuited in the thoroughly qualitative, concrete manner indicated in concluding the previous subsection. In the interest of formulating dimensional generation mathematically, we might be inclined to try to interpret terms like *zero density* and *infinite density* in a formal, purely algebraic fashion, as "well-defined points" in a nongeometric "function space," but this would serve only to perpetuate the classical intuition of continuity at a higher level of abstraction.

5. In discussing the characteristics of classical space, Capek (1961) asserted that "juxtaposition is the very essence of spatiality" (p. 19). See also Heidegger's (1962) characterization of classical space as sheer externality, the " 'outside-of-one-another' of the multiplicity of points" (p. 481).

6. The liberal use of quotation marks in portraying dimensional generation in quantitative terms is necessitated by the fact that such a portrayal does not do full justice to the thoroughly *qualitative* character of the process. A related observation was made in note 4, and the matter will be taken up again in section 3.7, where dimensional transformation will be characterized in a quasi-mathematical fashion, through metaphorical use of nonreal numbers.

7. For a somewhat different usage of this term, see Capek 1961.

8. It might be said in this regard that the contribution of Einsteinian relativity was to recognize that space and time indeed are interdependent. However, Einstein was not prepared to relinquish classical continuity as his guiding intuition and, therefore, did not carry his formulation of spatiotemporal interdependence far enough. See the reference to relativity on p. 67, and subsequent discussions of the role played by time in dimensional generation.

9. In figure 5.5a, it is certainly possible to imagine an "opposite side" of the plane, and this would imply perspective. *Neo-intuitive* imagination is required here. Opposite sides must be regarded as "separated" by a closed infinity, i.e., not yet differentiable, primordially identified.

10. It is easy to demonstrate that diametrically opposed perspectives of a three-dimensional object must be mirror images of each other; see Rucker's (1984, p. 46) whimsical confirmation of this for the forms of figure 5.5b.

11. Note that table 5.2 constitutes a purely qualitative expansion of the quasi-mathematical summary given in table 5.1. "Mathematization"

of table 5.2 would require introduction of "hypernumbers" beyond i and ϵ by which the deeper orders of primordiality metaphorically could be expressed.

12. These quotation marks speak to an important and difficult issue whose adequate treatment would require far more space than remains. The generalized Moebius concept implies that the primordiality of dimensional entities does not merely vary relative to changes in epoch but is inherent in the entities themselves. The generative process by which a lower-dimensional entity develops is such that the entity actually is left in a qualitatively less developed state than are higher-dimensional entities. Thus, the line is not just extended in one dimension fewer than the plane—it is less extensive in *quality*, intrinsically more intensive. This qualitative primordiality of an entity (neither brought out in section 3 nor reflected in table 5.2) is what is hidden in epochs beyond that of its generation. In our epoch, for example, the manifestation of the line and plane as mere extensive components of three-space completely masks their intrinsic qualities, and the even more intensive quality of the point shows up as but an absence of extension. The point, as the ultimate primordial entity, would have been left entirely ungenerated in its own epoch. Accordingly, we may regard epoch A as the "null epoch" of dimensional generation.

13. It may have occurred to the reader familiar with contemporary cosmology that it might be possible to establish a relation between the notion of epochs of dimensional generation and that of "eras of inflation," in the so-called inflationary model of the universe. Alas, I must again resist the temptation to "expand."

14. See Bohm's (1965) discussion of the relationship between the concept of invariance and perceptual development.

CHAPTER 6 (1992)

1. Heideggerian reflection on history reveals that the Renaissance actually constituted only one of several watersheds in the ongoing transformation of *dasein*, of human being-in-the-world. However, in this essay I will not explicitly consider any of the others. See Heidegger, 1975, for an idea of the change that occurred at the dawn of Greek thinking, and Heidegger, 1977b, for an indication of the most recent transformation of *dasein*.

2. Heidegger gives Galileo's phrase "I think in my mind of something moveable" (1977a, p. 267) as a prominent example of the early working of this new sense of self.

[3. *In the final section of chapter 5, the historical discussion of perspective is given a somewhat different emphasis. Rather than suggesting that perspectival experience arose with the Renaissance,*

I propose that it became more detached and abstract at this time. But there is no essential contradiction here, since the hallmark of perspectival awareness indeed is abstractive detachment. We can say, accordingly, that for less detached, more participatory pre-Renaissance consciousness, perspective was manifested in a relatively incipient form. Thus, while it would not be accurate to claim that pre-Renaissance experience was utterly unperspectival, we certainly may say that it was substantially less perspectival than was Renaissance experience. In any case, this distinction is not explicitly developed in the present chapter.]

CHAPTER 8 (1977)

1. This is physicist Hermann Weyl's characterization of the physical inaccessibility of hypothesized gaps in the Euclidean continuum. See Capek 1961, p. 226.
2. The title of this section is a play on a manner of speech used by Wheeler (1962). In his attempt to describe physical reality in the purely geometric terms of general relativity, he employed such phrases as "matter without matter," "force without force," and so on.
[3. In this 1977 account, I did not deal with the important distinction between the global/topological structure of space and its intrinsic character. See chapter 5.]

CHAPTER 9 (1983)

1. This neologism occurred to me in connection with a paper by psychologist Sigmund Koch (1981) in which the pervasive "epistemopathy" of psychological science was described. Koch himself might have used the term *epistemotherapy*, had he believed the ailment was treatable.

CHAPTER 11 (1987)

[1. There is an important distinction between the conventional interpretation of the "fourth dimension" and my own. In conventional mathematics, "higher dimensionalities" are summoned into being by extrapolation from the known three-dimensionality of the objective physical world. This procedure of dimensional proliferation is an act of abstraction presupposing that the character of dimensionality itself is left unchanged, that there is no alteration of its simple continuity (see chapter 5, p. 97). On this view, the "fourth dimension" required to complete the Klein bottle essentially remains an extensive, external dimension, though this "higher space" is taken as "imaginary." By contrast, I am proposing that

we grasp the "fourth dimension" as the one that is intensively folded within us, involving our thoughts, feelings, sensations and intuitions, going right down to their roots, right through the core of subjectivity to encompass the psyche. *In recent years, I have discovered that my interpretation of dimensionality is related in important ways to the views of philosopher Jean Gebser. See Gebser's magnum opus,* The Ever-Present Origin *(1985).]*

CHAPTER 12 (1982)

1. In order to keep my presentation within manageable bounds, I will not systematically consider the undeniably important role played by *time*.
2. See philosopher Milic Capek's (1961) discussions of the relationship between indeterminism and discontinuity.
[3. As noted in chapter 5, there is an important distinction *between "fractal" and "fractional" dimensionality, at least in the way these two notions have been formulated. Although fractal transformations are indeed non-topological, basically, they have been treated in separation from the classical concept of topological dimension, leaving the latter intact, not directly challenging it. By contrast, the idea of fractional dimensionality has been developed so as to focus upon and articulate the underlying* interrelationship *between classical and nonclassical domains, and this has led to a more general theory in which classical dimension is effectively (i.e., intrinsically) subsumed as a special case of an essentially nonclassical world.]*

CHAPTER 14 (1984)

1. The defining characteristics of epsilon are:

$$\epsilon^2 = 1, \epsilon \neq \pm 1.$$

EPILOGUE

1. This notion is elaborated in a recent essay of mine (Rosen 1993) and, independently, in the work of philosopher/psychologist Eugene Gendlin (1991).
2. The need to make our "ever-present origin diaphanous" is the central theme in the work of cultural philosopher Jean Gebser (1985).
3. Thinkers like Gebser (1985) and Burrow (1926) spoke of our need to retract or reclaim our projections, in the sense of making conscious their pre- or unconscious source. And of course, the withdrawal of

projections is among the most salient principles in the psychothera-
peutic program of C. G. Jung.

4. The writings of feminist philosophers like Christa Wolf (1984), Luce
Irigaray (1985), and Helene Cixous (1980) are highly relevant to the
question at hand, and they have influenced me considerably in recent
years (see Rosen 1993).

BIBLIOGRAPHY

Alapack, R. J. *J. of Phenomenological Psych.*, 2, 1971, 27.

Anderson, A. *God in a Nutshell*. Quincy, Mass.: Squantum Press, 1981.

Atkins, K. R. *Physics Once Over-Lightly*. New York: Wiley, 1972.

Bardwell, S. *Fusion Energy Found. Newsl.*, 2, 1977, 4.

Barfield, O. *Saving of Appearances*. Middletown, Ct.: Wesleyan U. Press, 1988.

Beloff, J. *J. Religion and Psychical Research*, 8, 1985, 3.

Bergson, H. *Duration and Simultaneity*. Indianapolis: Bobbs-Merrill, 1966.

Bjelland, A. In D. R. Griffin (Ed.), *Physics and the Ultimate Significance of Time*. Albany, N.Y.: SUNY Press, 1986, 51–80.

Bohm, D. *Quantum Theory*. Englewood Cliffs, N. J.: Prentice-Hall, 1951.

———. *The Special Theory of Relativity*. New York: Benjamin, 1965 (Appendix).

———. *Found. Phys.*, 3, 1973, 139.

———. *Wholeness and the Implicate Order*. London: Routledge & Kegan Paul, 1980.

———. In D. Griffin (Ed.), *Physics and the Ultimate Significance of Time*. Albany, N.Y.: SUNY Press, 1986, 177–208.

281

Bohm, D. and Hiley, B. J. *Found. Phys.*, 5, 1975, 93.

Brain/Mind Bulletin, 2, 1977, 1.

Brier, B. In S. C. Thakur (Ed.), *Philosophy and Psychical Research*. London: Allen & Unwin, 1976, 46–58.

Burrow, T. *Brit. J. Medical Psychology*, VI (Part III), 1926, 209.

Cantor, G. *The Campaigner*, 9, 1976, 69.

Capek, M. *Philosophical Impact of Contemporary Physics*. New York: Van Nostrand, 1961.

Capra, F. *The Tao of Physics*. Berkeley: Shambhala, 1975.

Carter, B. *Phys. Rev.*, 141, 1966, 1242.

———. *Phys. Rev.*, 174, 1968, 1559.

Chari, C. T. K. *J. Parapsychology*, 38, 1974, 1.

Cixous, H. In E. Marks and I. de Courtivron (Eds.), *New French Feminisms*. Sussex: Harvester Press, 1980, 245–64.

Comfort, A. *Reality and Empathy*. Albany, N.Y.: SUNY Press, 1984.

Cremmer, E. and Scherk, J. *Nucl. Phys.*, B103, 1976, 399.

Dobbs, A. *Proc. Soc. for Psychical Research*, 54, 1965, 249.

Dunne, J. W. *An Experiment with Time*. New York: Macmillan, 1938.

Eddington, A. S. *Fundamental Theory*. London: Cambridge U. Press, 1946.

Eisenbud, J. In W. G. Roll (Ed.), *Proceedings of the Parapsychological Association 1966*. Durham, N. C.: Parapsychological Association, 1967 (no. 3), 63–79.

Escher, M. C. *The Graphic Work of M. C. Escher.* New York: Ballantine, 1971.

Flew, A. G. N. In P. A. Schilpp (Ed.), *The Philosophy of C. D. Broad.* New York: Tudor, 1959, 411–435.

von Franz, M. *Number and Time.* Evanston, Ill.: Northwestern U. Press, 1974.

————. *Quadrant,* 8, 1975, 33.

Frescura, F. A. M. and Hiley, B. J. *Found. Phys.,* 10, 1980, 7.

Fuller, B. and McMurrin, S. *A History of Philosophy* (3rd Edition). New York: Holt, 1957.

Gamow, G. *One Two Three. . . Infinity.* New York: Bantam, 1961.

Gardner, M. *The Ambidextrous Universe.* New York: Basic Books, 1964.

Gebser, J. *Main Currents in Modern Thought,* 32, 1975, 198.

————. *The Ever-Present Origin* (N. Barstad, trans.). Athens, Ohio: Ohio U. Press, 1985.

Gendlin, E. T. and Lemke, J. *Math. Modeling,* 4, 1983, 61.

————. In B. den Ouden and M. Moen (Eds.), *The Presence of Feeling in Thought.* New York: Peter Lang, 1991, 27.

Georgescu-Roegen, N. *The Entropy Law and the Economic Process.* Cambridge: Cambridge U. Press, 1971.

Graves, J. *The Conceptual Foundations of Contemporary Relativity Theory.* Cambridge, Mass.: MIT Press, 1971.

Griffin, D. R. *Physics and the Ultimate Significance of Time.* Albany, N.Y.: SUNY Press, 1986a.

————. In D. R. Griffin (Ed.), *Physics and the Ultimate Significance of Time.* Albany, N. Y.: SUNY Press, 1986b, 127–53.

————. *The Reenchantment of Science.* Albany, N.Y.: SUNY Press, 1988.

Grobstein, C. In H. Pattee (Ed.), *Hierarchy Theory.* New York: George Braziller, 1973, 29–48.

Haight, D. F. "Remembrance of Things Present: A Second Copernican Revolution in Consciousness." Paper presented to Association for Transpersonal Psychology, Asilomar, California, June, 1983.

Heidegger, M. *Being and Time.* (J. Macquarrie and E. Robinson, trans.). New York: Harper & Row, 1962.

————. *Early Greek Thinking.* New York: Harper & Row, 1975.

————. In D. Krell (Ed.), *Martin Heidegger: Basic Writings.* New York: Harper & Row, 1977a, 243–82.

————. In D. Krell (Ed.), *Martin Heidegger: Basic Writings.* New York: Harper & Row, 1977b, 369–92.

————. In W. Lovitt (trans.), *The Question Concerning Technology and Other Essays.* New York: Harper & Row, 1977c, 3–35.

————. In W. Lovitt (trans.), *The Question Concerning Technology and Other Essays.* New York: Harper & Row, 1977d, 36–52.

————. Transcript of 1976 *Der Spiegel* interview with Heidegger (D. Schendler, trans.). *Grad. Faculty Philos. J.*, 6, 1977e, 4.

Heisenberg, W. *Physics and Philosophy.* New York: Harper & Row, 1958.

Hooker, C. A. *Man/Environ. Syst.*, 12, 1982, 121.

Hopkins, J. *Philosophical Rev.*, 82, 1973, 3.

Hurewicz, W. and Wallman, H. *Dimension Theory.* Princeton, N. J.: Princeton U. Press, 1941.

Irigaray, L. *This Sex Which is Not One*. Ithaca, N. Y.: Cornell U. Press, 1985.

Jacobi, J. *The Psychology of C. G. Jung*. New Haven: Yale U. Press, 1962.

Jahn, R. G. *Proc. of IEEE, 70, 1982, 136*.

Jahn, R. and Dunne, B. *On the Quantum Mechanics of Consciousness*. Princeton, N. J.: Princeton U. School of Engineering/Applied Sciences, 1983 (Appendix B).

Jones, W. T. *A History of Western Philosophy*. New York: Harcourt, Brace & World, 1952.

Josephson, B. D. *Phys. Educ.*, 22, 1987, 15.

Jung, C. G. In *The Interpretation of Nature and the Psyche*. New York: Pantheon, 1955, 1–146.

―――. In *The Structure and Dynamics of the Psyche* (Collected Works, Vol. 8). Princeton, N.J.: Princeton U. Press, 1969, 67–91.

―――. *Man and His Symbols*. New York: Doubleday, 1964.

Kaluza, T. *Sitzungsber. Preuss. Akad. Wiss. Phys. Math.*, K1, 1921, 966.

Karush, W. *Crescent Dictionary of Mathematics*. New York: Macmillan, 1962.

Kerferd, J. In C. C. Gillespie (Ed.), *Dictionary of Scientific Biography, Vol. IV*. New York: Scribner's, 1971, 30–35.

Klein, O. *Z. Phys.*, 37, 1926, 895.

Kline, M. *Mathematics: The Loss of Certainty*. New York: Oxford U. Press, 1980.

Koch, S. *Amer. Psychologist*, 36, 1981, 257.

Koestler, A. *The Act of Creation.* New York: Macmillan, 1964.

―――. *The Roots of Coincidence.* New York: Vintage, 1973.

Kramer, E. E. *The Nature and Growth of Modern Mathematics.* New York: Hawthorn, 1970.

Krippner, S. In R. A. White and R. Broughton (Eds.), *Research in Parapsychology 1983.* Metuchen, N.J.: Scarecrow Press, 1984, 153–66.

Lachelier, J. In J. T. Wilde and W. Kimmel (Eds.), *The Search for Being.* New York: Noonday Press, 1962, 151–74.

Lao Tsu. *Tao Te Ching* (G. Feng and J. English, trans.). New York: Vintage, 1972.

Lautman, A. In F. Le Lionnais (Ed.), *Great Currents in Mathematical Thought, Vol. I.* New York: Dover, 1971, 44–56.

Lavery, D. *Re-Vision,* 6, 1983, 22.

LeShan, L. *From Newton to ESP.* Wellingborough, Northamptonshire: Turnstone Press, 1984.

Levin, D. M. *The Body's Recollection of Being.* London: Routledge & Kegan Paul, 1985.

Macquarrie, J. *Martin Heidegger.* Richmond, Va.: John Knox Press, 1968.

Mandel, S. *Dictionary of Science.* New York: Dell, 1969.

Mandelbrot, B. *Fractals.* San Francisco: Freeman, 1977.

Mattuck, R. D. In *Parascience Proceedings 1973–1977.* London: The Institute of Parascience, 1977, 291–308.

Menger, K. *Rice Inst. Pamphlet,* 27, 1940, 1.

Merleau-Ponty, M. *The Visible and the Invisible*. Evanston, Ill.: Northwestern U. Press, 1968.

Mitroff, I. I., and Kilman, R. H. *Methodological Approaches to Social Science*. San Francisco: Jossey-Bass, 1978.

Mundle, C. W. K. *Proc. Soc. Psychical Research*, 56, 1973, 1.

Muses, C. J. *Stud. Consciousness*, 1, 1968, 29.

—————. In *Proceedings of the Second International Congress on Psychotronic Research*. Monte Carlo: International Association for Psychotronic Research, 1975a, 26–29.

—————. *Math./Physical Corresp.*, 11, 1975b, 17.

—————. *Impact of Science on Society* (UNESCO), 27, 1977, 67.

—————. *Destiny and Control in Human Systems*. Boston: Kluwer-Nijhoff, 1985.

de Nicolas, A. *Four Dimensional Man*. Bangalore, India: Dharmaram College, 1971.

Ong, W. *Interfaces of the Word*. Ithaca, N.Y.: Cornell U. Press, 1977.

Osis, K. *J. Amer. Soc. Psychical Research*, 59, 1965, 22.

Pattee, H. In E. Laszlo (Ed.), *The Relevance of General Systems Theory*. New York: George Braziller, 1972, 31–42.

—————. In H. Pattee (Ed.), *Hierarchy Theory*. New York: George Braziller, 1973, 71–108.

Pearce, J. C. *The Crack in the Cosmic Egg*. New York: Pocket Books, 1973.

—————. *The Bond of Power*. New York: Dutton, 1981.

Pratt, J. G. *J. Amer. Soc. Psychical Research*, 68, 1974, 133.

Progoff, I. *Jung, Synchronicity and Human Destiny.* New York: Julian Press, 1973.

Reed, D. *Energy Unlimited,* Part III–C, 1980, 33.

Reiser, O. *Cosmic Humanism.* Cambridge, Mass.: Schenkman, 1966.

———. *Cosmic Humanism and World Unity.* New York: Gordon & Breach, 1975.

Restivo, S. *Social Problems,* 35, 1988, 206.

Re-Vision, 1, 1978, 1.

Rogo, D. S. *Parapsychology Rev.,* 8, 1977, 20.

Rosen, S. M. *Scientia,* 108, 1973, 789.

———. In *Parascience Proceedings 1973–1977.* London: The Institute of Parascience, 1976, 229–45.

———. *Parapsychology Rev.,* 10, 1979, 8.

———. *Man/Environ. Syst.,* 11, 1981, 150.

———. *J. Religion and Psychical Research,* 6, 1983, 118.

———. *J. Religion and Psychical Research,* 8, 1985, 13.

———. "Wholeness as the Body of Paradox." Unpublished manuscript, 1993.

Rothenberg, A. *Arch. Gen. Psychiatry,* 33, 1976, 17.

Rucker, R. *Geometry, Relativity and the Fourth Dimension.* New York: Dover, 1977.

———. *The Fourth Dimension.* Boston: Houghton Mifflin, 1984.

Russell, B. *The ABC of Relativity.* New York: Harper & Bros., 1925.

Sartre, J. P. *Existentialism and Human Emotions*. New York: The Wisdom Library, 1957.

Scherk, J. and Schwarz, J. *Phys. Lett.*, 57B, 1975, 463.

Schultz, D. F. *A History of Modern Psychology*. New York: Academic Press, 1981.

Schumacher, J. *Human Posture*. Albany, N.Y.: SUNY Press, 1989.

Snow, C. P. *The Two Cultures: A Second Look*. Cambridge: Cambridge U. Press, 1964.

Stambaugh, J. In Heidegger, M., *On Time and Being* (J. Stambaugh, trans.). New York: Harper & Row, 1972, vii–xi.

Stanford, R. *J. Amer. Soc. Psychical Research*, 72, 1978, 197.

Stapp, H. *Found. Phys.*, 9, 1979, 1.

———. In D. R. Griffin (Ed.), *Physics and the Ultimate Significance of Time*. Albany, N.Y.: SUNY Press, 1986, 264–70.

Steiner, R. *An Outline of Occult Science*. Valley Spring, N.Y.: Anthroposophic Press, 1972.

Stiskin, N. *The Looking-Glass God*. Tokyo: Autumn Press, 1972.

Suzuki, D. T. *The Zen Doctrine of No Mind*. London: Rider, 1969.

Taylor, J. G. *Black Holes*. New York: Random House, 1973.

Tipler, F. J. *Phys. Rev.*, D9, 1974, 2203.

Tobin, B., Sarfatti, J. and Wolf, F. *Space-Time and Beyond*. New York: Dutton, 1975.

Ullman, M., Krippner, S. and Vaughan, A. *Dream Telepathy*. Jefferson, N. C.: McFarland, 1989.

Varela, F. J. *Internl. J. Gen. Syst.*, 2, 1975, 5.

Walker, E. H. In W. G. Roll, R. L. Morris and J. D. Morris (Eds.), *Research in Parapsychology 1972.* Metuchen, N.J.: Scarecrow Press, 1973, 51–53.

Washburn, M. *The Ego and the Dynamic Ground.* Albany, N.Y.: SUNY Press, 1988.

Watts, A. *The Supreme Identity.* New York: Pantheon, 1950.

Weber, R. In R. S. Valle and R. von Eckartsberg (Eds.), *Metaphors of Consciousness.* New York: Plenum, 1981, 121–40.

———. *Re-Vision*, 5, 1982, 35.

Werner, H. *Comparative Psychology of Mental Development.* Chicago: Follet, 1948.

Weyl, H. *Space-Time-Matter.* New York: Dover, 1922.

———. In J. R. Newman (Ed.), *The World of Mathematics, Vol. I.* New York: Simon & Schuster, 1956, 671–724.

Wheeler, J. A. *Geometrodynamics.* New York: Academic Press, 1962.

White, R. A. *Internl. J. Parapsychology*, 2, 1960, 5.

———. *J. Amer. Soc. Psychical Research*, 70, 1976, 333.

———. *J. Religion and Psychical Research*, 6, 1983, 220.

———. In R. A. White and J. Solfvin (Eds.), *Research in Parapsychology 1984.* Metuchen, N. J.: Scarecrow Press, 1985, 166–90.

Whitehead, A. N. *Process and Reality.* New York: Free Press, 1978.

Whitehead, A. N. and Russell, B. *Principia Mathematica.* Cambridge: Cambridge U. Press, 1925–27.

Whiteman, J. H. M. *J. Amer. Soc. Psychical Research*, 67, 1973, 341.

Wilber, K. *The Spectrum of Consciousness*. Wheaton, Ill.: Quest, 1977.

———. *Re-Vision*, 3, 1980, 51.

———. *Up From Eden*. New York: Doubleday/Anchor, 1981.

Witten, E. *Nucl. Phys.*, B186, 1981, 412.

Wolf, C. *Cassandra* (J. van Heurck, trans.). London: Virago, 1984.

Woods, R. *Mystical Spirituality*. Chicago: Thomas More Press, 1981.

Wu, C. S., Ambler, E., Hayward, R. W., Hoppes, D. D. and Hudson, R. P. *Phys. Rev.*, 105, 1957, 1413.

Zukav, G. *The Dancing Wu Li Masters*. New York: Morrow, 1979.

CREDIT ACKNOWLEDGMENTS

The author gratefully acknowledges permission he received to present the following: an adaptation of Alan Anderson's table, "Views of Reality," in *God in a Nutshell*, Quincy, Mass.: Squantum Press, (c) 1981; an adaptation of Clifford Grobstein's illustration, "Hierarchical Order and Neogenesis," in *Hierarchy Theory* (Ed. Howard Pattee), New York: George Braziller, (c) 1973; *Print Gallery* (c) 1956, *Tower of Babel* (c) 1928, and *Other World* (c) 1947, three works by M. C. Escher/Cordon Art—Baarn—Holland (all rights reserved), with reproduction rights arranged courtesy of the Vorpal Galleries, New York (Soho)/San Francisco; passages from David Bohm's *Wholeness and the Implicate Order*, published by Routledge & Kegan Paul, London, (c) 1980; George Gamow's drawing of an "inside-out universe," in *One Two Three. . . Infinity*, published by Bantam Books, New York, 1961, copyright (c) held by Professor R. Igor Gamow.

The author also would like to acknowledge the following journals and books in which the essays collected here first appeared: *Main Currents in Modern Thought*, 31 (4), 1975, pp. 115–20 (chapter 1); *Scientia*, 110, 1975, pp. 539–49 (chapter 2); *Man/Environment Systems*, 10 (5–6), 1980, pp. 239–50 and 12 (1), 1982, pp. 9–18 (chapters 3 and 12); *Speculations in Science and Technology*, 6 (4), 1983, pp. 413–25 (chapter 4); *Foundations of Physics*, 18 (11), 1988, pp. 1093–1139 (chapter 5); *The Interrelationship between Mind and Matter*, edited by Beverly Rubik and published by the Center for Frontier Sciences, Temple University, Philadelphia, 1992, pp. 233–47 (chapter 6); *The Journal of Phenomenological Psychology*, 5, 1974, pp. 33–39 (chapter 7); *Future Science*, edited by John White and Stanley Krippner and published by Doubleday/Anchor, New York, 1977, pp. 132–55, with permission granted by the late Ruth Hagy Brod (chapter 8); *Parapsychology Review*, 14 (1), 1983, pp. 17–24 (chapter 9); *Research in Parapsychology 1984*, edited by R. A. White and G. F. Solfvin and published by the Scarecrow Press of Metuchen, N.J. (copyright held by the Parapsychological

Association, Inc.), 1985, pp. 138–51 (chapter 10); *Parapsychology, Philosophy and Religious Concepts*, edited by B. Shapin and published by the Parapsychology Foundation, New York, 1987, pp. 68–97 (chapter 11); *Theta*, 10(4), 1982, pp.74–78 and 11(1), 1983, pp. 2–8 (chapter 12); *Physics and the Ultimate Significance of Time*, edited by D. R. Griffin and published by SUNY Press, Albany, N.Y., 1986, pp. 219–30 (chapter 14).

Finally, the author gives thanks to Mark A. Lewental (College Laboratory Technician, Library/Media Services, The College of Staten Island) for preparing a number of the illustrations appearing in this book.

INDEX

Active imagination. *See under* Imagination

Actuality. *See under* Potentiality

Affective, the, as an evolutionary stage, 41

Alapack, R. J., 127

Algebra: Clifford, and nilpotence, 219; discussed in Bohm/Rosen correspondence, 223–24, 246, 249; and invariance, 261–62, 267; and linearity and implicate order (Bohm), 218, 267; natural generalization of, 60. *See also* Muses: on hypernumbers; Nilpotence

Amino acid(s): and asymmetry, 18; and the creation of life, 38

Anderson, Alan: on basic theological positions, 185–86; and depiction of panentheism, 186, 188

Archetype: and dimension (space and time), 220; number as, 165, 218, 220, 245; as poetic (artistic), 235, 237, 240, 245, 254; as symbol of Self, 164, 217; universal, xii, 165, 218, 221, 254

Aristotle, and the question of Being, 183

Art: external representation vs. embodiment in, 191–94; and homospatial perception (Rothenberg), 250; and poetry, 247–48, 249; and symbolic representation, 215–16, 219, 228, 245, 251; and symmetry, 18; unity of expression in (Escher), 191, 237. *See also under* Archetype; Bohm

Aspect. *See under* Part

Aspecthood, 13, 36. *See also under* Freedom

Asymmetry (chirality, handedness): of boundary elements of space, 95; and dimension, 9, 17; of life, 19, 22; reduction of, to symmetry, 96; secondary status of, 19, 108; of space, 27; and stages of dimensional generation, 96; of the world, 24–25. *See also* Electron. *See also under* Amino acid; Dialectical process

Atheism. *See under* Materialism

Atkins, K. R., 22

Automorphism (transformation of space), 21. *See also under* Spiral

Bardwell, Steven, on discreteness in quantum field theory, 53, 153

Barfield, Owen, 117, 121

Becoming: as change, 3; process of (Bohm), 215; requirements of the concept of, 5. *See also* Being and Becoming; Process: and time; Wholeness: higher-order. *See also under* Moebius-Klein system; Moebius surface